Springer Tracts in Modern Physics
Volume 132

W0245939

Springer-Verlag
Berlin Heidelberg GmbH

Springer Tracts in Modern Physics

Volumes 118–134 are listed at the end of the book

Covering reviews with emphasis on the fields of Elementary Particle Physics, Solid-State Physics, Complex Systems, and Fundamental Astrophysics

Manuscripts for publication should be addressed to the editor mainly responsible for the field concerned:

Gerhard Höhler
Institut für Theoretische Teilchenphysik
Universität Karlsruhe
Postfach 6980
D-76128 Karlsruhe
Germany
Fax: +49 (7 21) 37 07 26
Phone: +49 (7 21) 6 08 33 75
Email: hoehler@fphvax.physik.uni-karlsruhe.de

Johann Kühn
Institut für Theoretische Teilchenphysik
Universität Karlsruhe
Postfach 6980
D-76128 Karlsruhe
Germany
Fax: +49 (7 21) 37 07 26
Phone: +49 (7 21) 6 08 33 72
Email: johann.kuehn@physik.uni-karlsruhe.de

Thomas Müller
IEKP
Fakultät für Physik
Universität Karlsruhe
Postfach 6980
D-76128 Karlsruhe
Germany
Fax: +49 (7 21) 6 07 26 21
Phone: +49 (7 21) 6 08 35 24
Email: mullerth@vxcern.cern.ch

Roberto Peccei
Department of Physics
University of California, Los Angeles
405 Hilgard Avenue
Los Angeles, California 90024-1547
USA
Fax: +1 310 825 9368
Phone: +1 310 825 1042
Email: robertop@college.ucla.edu

Frank Steiner
Abteilung für Theoretische Physik
Universität Ulm
Albert-Einstein-Allee 11
D-89069 Ulm
Germany
Fax: +49 (7 31) 5 02 29 24
Phone: +49 (7 31) 5 02 29 10
Email: steiner@physik.uni-ulm.de

Joachim Trümper
Max-Planck-Institut
für Extraterrestrische Physik
Postfach 1603
D-85740 Garching
Germany
Fax: +49 (89) 32 99 35 69
Phone: +49 (89) 32 99 35 59
Email: jtrumper@mpe-garching.mpg.de

Peter Wölfle
Institut für Theorie
der Kondensierten Materie
Universität Karlsruhe
Postfach 69 80
D-76128 Karlsruhe
Germany
Fax: +49 (7 21) 69 81 50
Phone: +49 (7 21) 6 08 35 90/33 67
Email: woelfle@tkm.physik.uni-karlsruhe.de

B. M. Andreev
E. P. Magomedbekov
G. H. Sicking

Interaction
of Hydrogen Isotopes
with Transition Metals
and Intermetallic
Compounds

With 72 Figures and 37 Tables

Springer

Professor Dr. B. M. Andreev

Mendeleev-Chemical-Technological Institute
9 Miusskaya Square
125 190 Moscow A190
Russia

Professor Dr. E. P. Magomedbekov

Mendeleev-Chemical-Technological Institute
9 Miusskaya Square
125 190 Moscow A190
Russia

Professor Dr. G. H. Sicking

Verbundzentrum für Oberflächen-
und Mikrobereichsanalyse
der Universitäten Düsseldorf und Münster
Wilhelm-Klemm-Str. 10
D-48149 Münster
Germany

Library of Congress Cataloging-in-Publication Data

Andreev, B. M. (Boris Mikhaĭlovich)
    Interaction of hydrogen isotopes with transition-metals and
intermetallic compounds / B.M. Andreev, E.P. Magomedbekov, G.
Sicking.
        p.   cm.
    Includes bibliographical references and index.

    1. Hydrogen--Isotopes. 2. Transition metal compounds.
3. Intermetallic compounds. 4. Exchange reactions. 5. Isotope
separation.   I. Title.
QD181.H1A524   1996
546'.21--dc20   95-39013
                                                              CIP

Physics and Astronomy Classification Scheme (PACS):
23, 28, 64, 82, 82.20

ISBN 978-3-662-14843-3   ISBN 978-3-540-48675-6 (eBook)
DOI 10.1007/978-3-540-48675-6

Typesetting: Adam Leinz, Karlsruhe
Cover design: Springer-Verlag, Design & Production
SPIN: 10477673   56/3144-5 4 3 2 1 0 - Printed on acid-free paper

Preface

The reader may be misled into understanding the title of this book in terms of merely "Hydrogen" in "Transition Metals and Intermetallic Compounds". This, however, would not co-incide with our intention, which, in fact, is well-described by adding and emphasizing the word "Isotopes".

Thus, the three hydrogen isotopes, protium (^{1}H or H), deuterium (^{2}H or D), and tritium (^{3}H or T), are principals in this book: their behaviour and interaction (both thermodynamic and kinetic) with metal–hydrogen systems (including intermetallic-compound–hydrogen systems) is comprehensively described.

The idea to write this review arose in 1985 on the occasion of our visiting some research groups working in the field of metal–hydrogen systems. Though they were using heavy hydrogen isotopes in their experiments there was a considerable lack of information with respect to isotope effects in metal–hydrogen systems.

In the literature, the behaviour of isotopes is classified under positive (normal) and negative (anomalous) isotope effects. The amazing (and irritating) point with metal–hydrogen systems is that in some of them the isotope effect can change from positive to negative and vice versa. Moreover, the situation is complicated and overlapped by the fact that the size of the isotope effect depends on the concentration ratio of the two hydrogen isotopes in question. In the literature, the description of protium–tritium isotope effects in metal–hydrogen systems usually relate to trace amounts of tritium and the situation of high tritium fractions is not touched, even in studies where isotope effects are considered in terms of solving the problems of tritium recovery, tritium enrichment, or tritium concentration.

Little wonder that many numerical values concerning isotope effects in metal–hydrogen systems (e. g. isotope separation factors) cannot directly be compared with one another. It is hoped that this review will clearly explore most of the parameters influencing isotope effects in metal–hydrogen systems, so that experimentalists in this field are helped to define proper conditions for their experiments.

In addition, the book presents and elucidates the interrelations between thermodynamic and kinetic isotope effects, on the one hand, and between temperature, hydrogen content, isotopic composition, polydispersity, lattice and geometrical structure, and electronic structure on the other. Basically the reader is put into a position to re-calculate numerical values from the literature for the purpose of comparison. In fact, these re-calculations have been carried out for quite a number of published data on isotope effects in metal–hydrogen systems.

The organization of this book is such that the basics on metal–hydrogen systems are given in Chap. 2. The equilibrium isotope effects are dealt with in Chap. 3, and the kinetics of the isotope exchange reaction, including mass transfer in a column and also including granulated sorbents, in Chap. 4.

Chapter 5 is devoted to processes of periodic separation and of continuous counter-current separation. This chapter is based on the interrelationships and data given in the previous chapters. Chapter 5 ends with a design study of a separation plant able to regenerate a D–T mixture as it emergings from a TNR plasma chamber. The designed separation plant makes use of the isotope exchange reaction between the gas-phase and Pd-hydride, whereby only about 10 kg of Pd is needed.

We greatly acknowledge recurring support from the national science foundations of both countries. Thereby, it is worth mentioning that support was given for the first time as early as 1979, enabling an extended scientific stay of one of us (E.P.M.) at the university of Münster, a quite exceptional event in those days.

Moscow and Münster,   *B. M. Andreev*
January 1996   *E. P. Magomedbekov*
                                            G. H. Sicking

Contents

1. Introduction

The interaction of hydrogen with metals is a major field within physical chemistry, which can be subdivided into several separate trends. The study of hydrogen–transition-metal systems and compounds based on them (intermetallic compounds or IMC) is one of the most interesting areas.

Over the past 30 years a wealth of information has been accumulated on electronic structure, thermodynamic and kinetic characteristics of sorption, segregation phenomena and magnetic properties of metal hydride phases and IMC. Very important experimental methods are the following: nuclear magnetic resonance (NMR), neutron and Mössbauer spectroscopy, laser mass-spectrometry, electron spectrometry and so on. The results of numerous investigations testify to the future prospects of this field. These results are presented at traditional symposia ("Hydrogen in Metals" since 1968, "Properties and Applications of Metal Hydrides" since 1977) and in several collected volumes within the series Topics in Applied Physics [1.1–3].

The essential feature of hydrogen–d-element systems consists in the formation of chemical bonds between hydrogen and metal atoms. Hence the study of the interaction of hydride-forming metals and IMC with heavy hydrogen isotopes – deuterium and tritium – opens up new possibilities for investigating hydrogen behaviour on surfaces and in a solid matrix. Non-standard and sometimes incorrectly explained physical effects (e. g., "cold thermonuclear synthesis") make these systems very attractive both in terms of theory and for practical application. The latter in particular, is related to the fact that in heavy-hydrogen-isotope hydride phases of transition metals or IMC one observes large isotope effects, which can be used to separate hydrogen isotopes. Many problems arising in the nuclear-fuel cycle can be solved by using hydride-forming metals and IMC; for example, recovery, purification, and storage of heavy hydrogen isotopes in tritium containing systems in thermonuclear reactors and the extraction and localization of tritium in heavy-water nuclear reactors.

The available information on the interaction of heavy hydrogen isotopes with transition metals and IMC is incomplete and unsystematic. Nonetheless, it enables one to consider a diversity of theoretical questions.

Up to now the literature contains no monograph elucidating theoretical aspects of the interaction of heavy hydrogen isotopes with metals and IMC and including systematic material on phases, isotope equilibrium and kinetics of hydrogen-isotope exchange in systems with hydride phases.

This work aims to fill the gap. During its preparation the authors used not only results known from the literature but also, and primarily, the results of their own theoretical and experimental investigations performed over the last 20 years.

In the second chapter, where the thermodynamics of hydrogen isotope interaction with hydride-forming metals and IMC is considered, the differences in phase equilibrium for protium, deuterium, and tritium with metals and IMC are presented and analyzed. These result in the thermodynamic isotope effect and are of interest in terms of the above-mentioned applications.

The third chapter is devoted to isotope equilibrium with hydride phases. It involves both theoretical questions of the thermodynamics of hydrogen-isotope exchange and integration of published experimental data and of isotope effects studied by the present authors.

The fourth chapter considers the kinetics of the isotope exchange reaction of hydrogen with hydride phases, describes the deviations discovered by the authors from the commonly used kinetic equation, determined by polydispersity of solid phases of metals and IMC, and by segregation effects on their surfaces. A major part of Chap. 4 is devoted to discussing the efficiency of mass-transfer in columns with IMC.

Chapter 5 reviews the literature on counter-current separation processes (both periodic and continuous) of hydrogen isotopes on Pd, V, and IMC. The authors then consider the most effective – in their opinion – counter-current separation processes in a sectioned column with an immobile bed of hydrogen sorbent. Based on the presented data on separation factors (Chap. 3) and kinetics of interphase isotope exchange (Chap. 4) they discuss questions of optimization of separation plants and ways of practical application of hydride phase systems (foremost based on palladium) for solving tasks of isotope purification in the heavy-water moderator in nuclear reactors and for creating tritium cycles and radiation safety systems for thermal nuclear reactors.

2. Thermodynamics of the Interaction of Hydrogen Isotopes with Metals and Intermetallic Compounds

2.1 Theory of Phase Equilibrium in Hydrogen–Metal Systems

Let us consider dilute solutions of hydrogen in transition metals, in which formation of new structure does not occur and hydrogen can be considered as randomly distributed at possible positions (interstices) in the crystal lattice.

For this range one can write

$$\Delta \overline{G}_H = RT \ln P_{H_2}^{1/2} = \Delta \overline{H}_H - T \Delta \overline{S}_H , \tag{2.1}$$

where P_{H_2} is the hydrogen equilibrium pressure (in bar), $\Delta \overline{G}_H$ is the change of the molar isobaric-isothermal potential (change in GIBBS' free energy), and $\Delta \overline{H}$ and $\Delta \overline{S}$ are the changes of the partial molar enthalpy and entropy during the process of hydrogen solution in metal. Since $\Delta \overline{G}_H = \overline{G}_H - \frac{1}{2} G_{H_2}^0$, the remaining terms of (2.1) can be defined as $\Delta \overline{H}_H = \overline{H}_H - \frac{1}{2} H_{H_2}^0$, $\Delta \overline{S}_H = \overline{S}_H - \frac{1}{2} S_{H_2}^0$.

The influence of imperfections (H–H interaction, dilatation of interstices, change of the Fermi level due to the electrons of dissolved hydrogen) may be ignored at low hydrogen concentrations [2.1]. Thus (2.1) can be written as

$$\Delta \overline{G}_H = \Delta \overline{H}_H^0 - T \left(\overline{S}_H^0 + \overline{S}_H^{ci} - \frac{1}{2} S_{H_2}^0 \right) = \Delta \overline{G}_H^0 - T \overline{S}_H^{ci} , \tag{2.2}$$

where $\Delta \overline{H}_H^0$ is the partial molar enthalpy of solution. $\Delta \overline{S}_H^0$ is unconfigurational partial molar entropy of hydrogen solution. $\overline{S}_H^{ci}$ is the ideal configurational entropy, defined in accordance with equation

$$\overline{S}_H^{ci} = -R \ln \frac{n/n_s}{1 - n/n_s} , \tag{2.3}$$

where n and n_s are the ratios of the number of hydrogen atoms to the number of metal atoms, n_s corresponding to the geometrically maximum possible number. Using (2.1–3) one can derive for $n \to 0$ and $1 - n/n_s \to 1$

$$\left(P_{H_2} \right)^{1/2} = \frac{n}{n_s} \exp \left[\frac{\Delta \overline{H}_H^0}{RT} - \frac{\Delta \overline{S}_H^0}{R} \right] = K_s n . \tag{2.4}$$

This formula is equivalent to Sieverts *law*.

From the dependence $P^{1/2}$ on n, the Sieverts constant K_s can be determined and from the temperature dependence $\ln K$ on $1/T$ values $\Delta \overline{H}_H^0$ and $\Delta \overline{S}_H^0$ can be found. In this case values of n_s must be known for the particular metal. The values of n_s depend on the metal structure and the type of interstices occupied by hydrogen atoms.

In most face centered cubic metals hydrogen occupies octahedral positions and the number of such positions per atom is equal to 1, in hexagonal close packed and body centered cubic metals, hydrogen occupies tetrahedral positions and n_s is equal to 2 and 6, respectively. These values are found from neutron diffraction experiments on Pd, Ni, Ta, Zr, Hf, and Ti hydrides [2.2–5]. A review of the data on hydrogen solubility in some transition metals is represented in [2.1, 6]. Using the experimental values of ω_H obtained by inelastic neutron scattering spectroscopy (INSS) the authors have calculated the entropy of solution $\Delta \bar{S}_H^0$ using the three-dimensional harmonic oscillator model. The calculated values for the metals mentioned above agree with the experimental ones with an accuracy of $4\,\mathrm{J\,mol^{-1}\,K^{-1}}$. This supports the validity of this model describing the behaviour of hydrogen atoms in the crystal lattice of metals.

At higher hydrogen pressures some transition metals exhibit a break of miscibility in which both phases (α and β) coexist and are in equilibrium with hydrogen. At temperatures lower then the critical one concentration–pressure isotherms have a plateau at hydrogen concentrations from n_α to n_β. The transition from the α-phase to the β-phase is often said to be hydride-formation. It can be written in the form of a reaction

$$\frac{2}{n_\beta - n_\alpha} \mathrm{MeH}_{n\alpha} + \mathrm{H}_2 = \frac{2}{n_\beta - n_\alpha} \mathrm{MeH}_{n\beta}. \tag{2.5}$$

To interpret the P-C-T diagram the van't Hoff equation is ordinarily used in the form

$$\ln P_{\mathrm{H}_2} = \frac{\Delta H_{\alpha-\beta}}{RT} - \frac{\Delta S_{\alpha-\beta}}{R}, \tag{2.6}$$

where $\Delta H_{\alpha-\beta}$ and $\Delta S_{\alpha-\beta}$ are the changes of enthalpy and entropy during hydride-formation (2.5) per mole of H_2. They can be expressed as

$$\Delta H_{\alpha-\beta} = 2 \frac{H_\beta - H_\alpha}{n_\beta - n_\alpha} - \bar{H}_{\mathrm{H}_2} \tag{2.7}$$

$$-\Delta S_{\alpha-\beta} = 2 \frac{S_\beta - S_\alpha}{n_\beta - n_\alpha} - \bar{S}_{\mathrm{H}_2}. \tag{2.8}$$

Assuming that $S_\alpha \approx S_\beta$ and $\bar{S}_{\mathrm{H}_2} = S_{\mathrm{H}_2}^0 - R \ln P_{\mathrm{H}_2}$, i.e., only slight entropy changes, the equilibrium condition (2.6) can be written in the form

$$\ln P_{\mathrm{H}_2} = \frac{\Delta H_{\alpha-\beta}}{RT} - \frac{S_{\mathrm{H}_2}^0}{R}, \tag{2.9}$$

where $S_{\mathrm{H}_2}^0$ is equal to $130.8\,\mathrm{J\,mol^{-1}\,K^{-1}}$. Thus the entropy term can be approximately equal for many of the metals.

Since the hydride-formation process is accompanied by dissociation of hydrogen into atoms, (2.9) must be written as

$$\ln\left(P_{H_2}^{1/2}\right) = \frac{\Delta H_{\alpha-\beta}}{RT} - \frac{S_{H_2}^0}{2R}.$$

(2.10)

In this case $\Delta H_{\alpha-\beta}$ is related to one g-atom of H. Equation (2.6) is appropriate in cases where $\Delta H_{\alpha-\beta}$ and $\Delta S_{\alpha-\beta}$ do not depend on temperature. It is shown in [2.7] that in the temperature range 200–700 K the value $\Delta S_{\alpha-\beta}$ remains constant for those metals in which hydrogen is located in tetrahedral interstices ($\theta_H = \hbar\omega_H/k$ at 1000–1100 K) ($\hbar$ means $h/2\pi$).

Analysis of $\Delta H_{\alpha-\beta}$ change using the Einstein oscillator model for describing the behaviour of hydrogen atoms in the crystal lattice of metals is presented in [2.8].

The value $\Delta H_{\alpha-\beta}$ is related to the enthalpy of hydride-formation under standard conditions by the equation

$$\Delta H_{\alpha-\beta}^0 = \Delta H_{\alpha-\beta} - \overline{V}_H(p - p_0)$$
$$- 3R\theta_H \left\{ \left[\exp\left(\frac{\theta_H}{T}\right) - 1 \right]^{-1} - \left[\exp\left(\frac{\theta_H}{T_0}\right) - 1 \right]^{-1} \right\}$$
$$+ \tfrac{1}{2}\left(\overline{H}_{H_2} - \overline{H}_{H_2}^0 \right).$$

(2.11)

Here and below $\Delta H_{\alpha-\beta}$ is referred to 1 g-atom of hydrogen. In (2.11) $\overline{V}_H$ is the volume occupied by 1 g-atom of hydrogen in the crystal lattice.

Considering the assumptions made for (2.9), namely, $\Delta H_{\alpha-\beta} = -1/2TS_{H_2}$, it can be found, introducing the chemical potential for the plateau, that

$$\Delta H_{\alpha-\beta}^0 = \tfrac{1}{2}\mu_{H_2} - \overline{V}_H(P - P^0)$$
$$- 3R\theta_H \left\{ \left[\exp\left(\frac{\theta_H}{T}\right) - 1 \right]^{-1} - \left[\exp\left(\frac{\theta_H}{T_0}\right) - 1 \right]^{-1} \right\}$$
$$- \tfrac{1}{2}\overline{H}_{H_2}^0.$$

(2.12)

Substituting $P_0 = 1$ bar, $T_0 = 298$ K, $H_{H_2}^0 = 8.45$ kJ/(mol H_2), and $\overline{V}_H = 1.7$ cm^3/g-atom H (typical value for some hydrides) in (2.12) and using the value of μ_{H_2} given in [2.8] at a characteristic temperature averaged for tetrahedral (1100 K) and octahedral (600 K) positions $\theta_H = 850$ K the authors showed that in the temperature range from 200 to 1000 K at pressures $P \leq 1000$ bar the dependence $\ln P_{H_2} - 1/T$ is well described by a straight line.

2.2 Phase Equilibrium
in Heavy Hydrogen Isotope–Metal (IMC) Systems

Let us use the thermodynamic consideration presented in Sect. 2.1 for an analysis of the state of heavy hydrogen isotopes – deuterium and tritium – at phase equilibrium with deuterides and tritides of metals and IMC.

At phase equilibrium the values of the equilibrium pressures P_{H_2}, P_{D_2}, and P_{T_2} at the same concentration n_H, n_D, and n_T in hydrides, deuterides, and tritides, respectively, are the most essential parameters.

For diluted solutions, in accordance with (2.4), for the ratio of pressures of light isotope A_2 to the heavy one B_2, one can derive

$$\ln \left(\frac{P_{A_2}}{P_{B_2}} \right)^{1/2} = \frac{\Delta \overline{H}_A^0 - \Delta \overline{H}_B^0}{RT} - \frac{\Delta \overline{S}_A^0 - \Delta \overline{S}_B^0}{R}. \tag{2.13}$$

The difference of enthalpies and entropies of solution of the two isotopes can be written

$$\Delta \overline{H}_A^0 - \Delta \overline{H}_B^0 = \overline{H}_A - \overline{H}_B - \tfrac{1}{2} \left(H_{A_2}^0 - H_{B_2}^0 \right) \tag{2.14}$$

$$\Delta \overline{S}_A^0 - \Delta \overline{S}_B^0 = \overline{S}_A - \overline{S}_B - \tfrac{1}{2} \left(S_{A_2}^0 - S_{B_2}^0 \right) \tag{2.15}$$

Now it is possible to write (2.13) in the form

$$\ln \left(\frac{P_{A_2}}{P_{B_2}} \right)^{1/2} = \frac{\overline{H}_A - \overline{H}_B - \tfrac{1}{2} \left(H_{A_2}^0 - H_{B_2}^0 \right)}{RT}$$
$$- \frac{\overline{S}_A - \overline{S}_B - \tfrac{1}{2} \left(S_{A_2}^0 - S_{B_2}^0 \right)}{R} \tag{2.16}$$

Let us transform (2.16) so that the first term contains the thermodynamic characteristics of the gas phase and the second term those of the solid phase:

$$\ln \left(\frac{P_{A_2}}{P_{B_2}} \right)^{1/2} = \frac{H_{B_2}^0 - H_{A_2}^0}{2RT} - \frac{S_{B_2}^0 - S_{A_2}^0}{2R} + \frac{\overline{H}_A - \overline{H}_B}{RT} - \frac{\overline{S}_A - \overline{S}_B}{R}. \tag{2.17}$$

The first term on the right of (2.17) accounts for the difference in energy states of the two isotope modifications of molecular hydrogen, the second term reflects the difference in energy states of the two isotopes when dissolved in the crystal lattice of metals.

The chemical potentials of the atoms in interstices can be found from the relationship $\mu_A = \overline{H}_A - T\overline{S}_A$. The same relationship holds for the thermodynamic parameters of molecular hydrogen: $\mu_{A_2} = H_{A_2} - TS_{A_2}$. In terms of these relations and using (2.17) one can obtain the following expression

$$RT \ln \left(\frac{P_{A_2}}{P_{B_2}} \right)^{1/2} = \frac{1}{2} \left(\mu_{B_2} - \mu_{A_2} \right) + \left(\mu_A - \mu_B \right). \tag{2.18}$$

As was shown above, the energy state of hydrogen atoms in the lattice of a metal is adequately described by the Einstein model. Ignoring the distinctions in components of chemical potentials μ_A and μ_B determined by the different energy of the crystal lattice deformation and by anharmonicity of oscillations of A and B in interstices (as was shown in [2.9] these two effects partially compensate each other), Eq. (2.18) can be written

$$\ln(P_{A_2}/P_{B_2})^{1/2} = \frac{\mu_{B_2} - \mu_{A_2}}{2RT} + \frac{3(\theta_A - \theta_B)}{2T}$$

$$+ 3\ln\left[\frac{1 - \exp(-\theta_A/T)}{1 - \exp(-\theta_B/T)}\right]. \tag{2.19}$$

Since the Einstein model anticipates the relation $\theta_A = \theta_B(m_B/m_A)^{1/2}$, introducing the reduced temperature $u = \theta/T$ one obtains

$$\ln\left(\frac{P_{A_2}}{P_{B_2}}\right)^{1/2} = \frac{\mu_{B_2} - \mu_{A_2}}{2RT} + 3\ln\left[\exp\left(\frac{u_A}{2}\right) - \exp\left(\frac{u_A}{2(m_B/m_A)^{1/2}}\right)\right]$$

$$+ 3\ln\left[\frac{1 - \exp(-u_A)}{1 - \exp(-u_A/(m_B/m_A)^{1/2})}\right] \tag{2.20}$$

The second and third term on the right of (2.20) can be written with the use of hyperbolic sine $\sinh = [1 - \exp(-u)]/2\exp(-u/2)$. Thus we obtain the following expression

$$\ln\left(\frac{P_{A_2}}{P_{B_2}}\right)^{1/2} = \frac{\mu_{B_2} - \mu_{A_2}}{2RT} + 3\ln\left[\frac{\sinh(u_A/2)}{\sinh(u_A/2(m_B/m_A)^{1/2})}\right]. \tag{2.21}$$

For the transition range given in (2.5), using an analogous approach and (2.10), one can derive, in terms of the assumptions made above (Sect. 2.1), the following expression:

$$\ln\left(\frac{P_{A_2}}{P_{B_2}}\right)^{1/2} = \frac{\Delta H^A_{\alpha-\beta} - \Delta H^B_{\alpha-\beta}}{RT} - \frac{S^0_{A_2} - S^0_{B_2}}{2R}. \tag{2.22}$$

The second term on the right of this equation reflects the difference in entropies of isotope modifications of gaseous hydrogen and can be determined by tabulated data (for example, $(S^0_{T_2} - S^0_{D_2})/2R = -0.50381$). Let us represent the right-hand side of (2.22) in the form of two terms related to the gas and the solid phases. For this purpose we draw on (2.7, 8) divided by 2, in which H_α, H_β, and S_α, S_β are partial molar enthalpies and entropies of hydrides MeH_{n_α} and MeH_{n_β}, respectively. Using the Gibbs–Duhem equations they can be written

$$S_i = n_i\overline{S}^H_i + \overline{S}^{Me}_i, \tag{2.23}$$

$$H_i = n_i\overline{H}^H_i + \overline{H}^{Me}_i. \tag{2.24}$$

Considering that for any pair of hydrogen isotopes $(n_\beta - n_\alpha)^A = (n_\beta - n_\alpha)^B$, $[\overline{H}^{Me}_i]^A = [\overline{H}^{Me}_i]^B$, $[\overline{S}^{Me}_i]^A = [\overline{S}^{Me}_i]^B$, and $n_\beta \gg n_\alpha$ for the ratio of pressures of light isotope A_2 to heavy one B_2 in the transition range, the following expression similar to (2.17) can be derived

$$\ln\left(\frac{P_{A_2}}{P_{B_2}}\right)^{1/2} = \left(\frac{H^0_{B_2} - H^0_{A_2}}{2RT} - \frac{S^0_{B_2} - S^0_{A_2}}{2R}\right)$$
$$+ \left(\frac{H^A_\beta - H^B_\beta}{RT} - \frac{S^A_\beta - S^B_\beta}{R}\right). \tag{2.25}$$

Values of H^A_β, H^B_β, S^A_β, and S^B_β can be estimated using the lattice-gas model [2.8] which enables (2.25) to be represented similarly to (2.21); the latter differs in that values of u_A are connected with the frequencies of local modes of hydrogen atoms in metal interstices in the hydride β-phase.

In the case $u^\alpha_A > u^\beta_A$ as for Pd [2.9, 10], the value of $\ln P_{A_2}/P_{B_2}$ decreases in the β-phase.

For the majority of metals (Chap. 3) $u^\alpha_A \approx u^\beta_A$ and values of $\ln P_{A_2}/P_{B_2}$ are approximately equal for α- and β-phases.

Experimental data on phase equilibrium of hydrogen isotopes with metals (IMC) are available only for a few systems and mainly for deuterium.

The difficulties associated with carrying out experiments with radioactive tritium and, in particular, with a great amount of tritium of about 500–1000 Ci, required to study the equilibrium in the β-phase, have constrained the number of studies on pure tritium sorption. Practically all the work with pure tritium is performed in Jülich.

Now the Pd–H$_2$ system is the sole system, for which repeatable data on phase equilibrium of all the isotopes are available. The results of investigations of hydrogen-isotope–palladium systems are compiled in [2.10, 11].

Figure 2.1 shows the isotherms of sorption of all hydrogen isotopes. As is evident from Fig. 2.1 in the studied temperature range $P_{T_2} > P_{D_2} > P_{H_2}$ at the same content and $n_T < n_D < n_H$ at the same pressure. This points to concentration of heavy isotopes in the gas phase. Using the data obtained in [2.11–13] it is possible to analyze the influence of hysteresis for the protium–deuterium pair on the isotope effect, since for tritium the data are available either only for sorption isotherms [2.13] or only for desorption isotherms [2.11]. This hinders the comparison although values $P^A_{T_2}/P^A_{H_2} = 12.23$ and $P^D_{T_2}/P^D_{H_2} = 9.04$ at 298 K are presented in [2.13]. The absence of experimental data on sorption isotherms of tritium prohibits an analysis of the dependence of this parameter on experimental conditions.

Table 2.1 presents the calculated values of the isotope effect P_{D_2}/P_{H_2} determined from the sorption and desorption isotherms, and the values of the hysteresis for protium and deuterium at different temperatures.

The following conclusions can be drawn from Table 2.1: Firstly, the value of the hysteresis in the Pd–D$_2$ system is higher than that in the Pd–H$_2$ system; secondly, the isotope effect determined by the sorption isotherms is higher than that determined by the desorption isotherms, with the exception of the data at 298 K [2.13]. This difference decreases with increasing temperature and at 383–393 K it lies within the limits of experimental error.

From the above thermodynamic consideration and the lattice-gas model it is impossible to explain the discrepancies in the values of the isotope effects, since

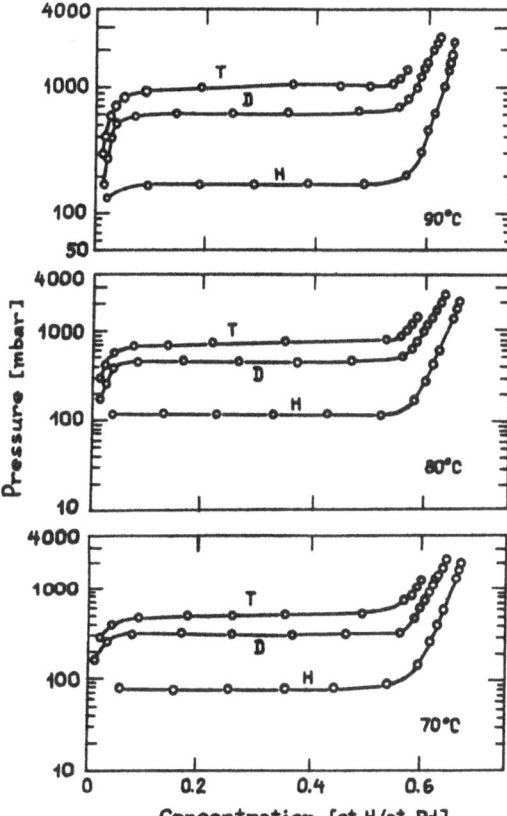

Fig. 2.1. Isotherms of hydrogen isotope desorption on palladium [2.11]

the model does not take into account the distinction between H and D with respect to the different energy of lattice deformation resulting from proton and deuteron implantation at an interstitial site. The root-mean-square shifts in the β-phase determined by neutron scattering are found to be equal to 0.23 Å for H and 0.20 Å for D [2.14], i.e. deuterium atoms require less volume. This fact can lead to a different value of the hysteresis related to the lattice deformation.

However, for other systems this distinction can be less essential, for instance, for LaNi$_5$, or can be directly opposed to that in Pd-H$_2$(D$_2$), as is the case for TiMn$_{1.5}$. Calculation of the values in Table 2.1 is performed using the sorption–desorption isotherms of protium and deuterium presented in [2.15, 16] for LaNi$_5$ and TiMn$_{1.5}$, respectively.

The published data on the interaction of hydrogen isotopes with metals and IMC – as far as they are familiar to the authors – are summarized in Tables 2.2 and 2.3.

The thermodynamic parameters of sorption of hydrogen by metals and IMC are included in Tables 2.2, 3. As is noted above the difference in the heat of formation ΔH_j of the corresponding pair of isotopes has a decisive effect on the value of the

Table 2.1. Temperature dependence of isotope effect and hysteresis

Me	T/K	$P_{D_2}A/P_{H_2}A$	$P_{D_2}D/P_{H_2}D$	$\ln(P_{H_2}A/P_{H_2}D)$	$\ln(P_{D_2}A/P_{D_2}D)$
Pd	298	5.11	5.14	0.89	0.88
	333	4.34	4.00	0.50	0.59
	343	4.28	3.95	0.52	0.60
	353	4.03	3.83	0.53	0.58
	363	3.81	3.64	0.52	0.57
	373	3.65	3.54	0.51	0.54
	383	3.53	3.49	0.50	0.51
	393	3.40	3.42	0.49	0.49
LaNi$_5$	273	0.94	0.86	0.24	0.31
	323	1.20	1.08	0.29	0.39
	388	1.19	1.15	0.31	0.34
TiMn$_{1.5}$	195	0.18	0.23	1.30	1.05
	228	0.26	0.34	0.99	0.69
	273	0.39	0.46	0.57	0.40

Table 2.2. Thermodynamic characteristics of hydrogen isotopes dissolving in metals and IMC

Me (IMC)	$-\Delta \overline{H}_H^0$ kJ/mol	$-\Delta \overline{H}_D^0$ $(-\Delta \overline{H}_T^0)$ kJ/mol	$-\Delta \overline{S}_H^0$ J/mol K	$-\Delta \overline{S}_D^0$ $(-\Delta \overline{S}_T^0)$ J/mol K	Reference
Pd	19.3	15.8	107.1	106.2	[2.17]
	18.8	15.7	105.0	105.5	[2.18]
	–	(13.8)	–	(103.8)	[2.19]
V	54.2	(63.4)	135.4	(152.4)	[2.20]
	63.4	65.6	149.0	152.6	[2.21]
	60.4	63.2	145.4	150.2	[2.22]
Ta	72.0	79.2	100.0	112.0	[2.23]
Ti	88.8	84.6	54.5	74.6	[2.24]
	89.7	83.0	–	–	[2.25]
Zr	103.9	103.9	99.6	–	[2.26]
	116.5	121.5	–	–	[2.27]
Ni	71.8	79.7	68.0	72.2	[2.28]
La	154.2	150.8 (142.5)	–	–	[2.29]
TiCr$_{1.8}$	70.0	75.3	144.5	158.8	[2.30]
TiMo	86.9	97.5	–	–	[2.31]

values in parenthesis relate to tritium

Table 2.3. Thermodynamic characteristics of hydrogen isotopes sorption by metals and IMC

Me (IMC)	$-\Delta H_H$ kJ/mol	$-\Delta H_D$ $(-\Delta H_T)$ kJ/mol	$-\Delta S_H$ J/mol K	$-\Delta S_D$ $(-\Delta S_T)$ J/mol K	Reference
Pd	39.0	37.2	91.2	97.8	[2.17]
	39.0	35.4 (33.3)	92.5	93.4 (91.7)	[2.17]
V	40.2	50.3	140.8	164.3	[2.32]
Ni	40.2	41.1	131.6	132.4	[2.32]
Ta	96.2	100.0	–	–	[2.23]
U	86.8	88.8 (90.7)	116.2	116.2 (116.2)	[2.23]
	125.3	132.0	179.3	196.1	[2.34]
La	199.0	203.2 (206.6)	142.5	148.3 (153.4)	[2.29]
LaNi$_5$	30.9	35.2	109.2	122.9	[2.15]
	29.0	33.5 (34.4)*	108.0	125.0 (130.0)*	[2.35]
SmCo$_5$	30.2	33.1 (35.1)*	113.1	121.0 (130.0)*	[2.35]
LaNi$_4$Cu	32.0	36.0 (38.0)*	112.0	126.0 (130.0)*	[2.36]
LaNi$_4$Cr	33.0	38.0 (39.0)*	112.0	126.0 (134.0)*	[2.36]
LaNi$_3$Cu$_2$	31.0	38.0 (39.0)*	96.0	142.0 (146.0)*	[2.36]
TiMn$_{1.5}$	27.9	31.2	112.2	119.0	[2.16]
TiMn$_{1.4}$Ni$_{0.1}$	28.9	31.6	109.5	116.1	[2.37]
CaNi$_5$	33.5	33.5	–	–	[2.38]
ZrCo	89.8	95.9	148.0	163.0	[2.39]
	116.4	84.5	–	–	[2.40]
ZrNi	96.3	71.5	–	–	[2.41]
	47.3	48.7	–	–	[2.40]

values in parenthesis relate to tritium   * calculated values

hydrogen-isotope effect at $|\Delta H_D| - |\Delta H_H| > 0$. It means that deuterium (the heavy isotope) is concentrated in the solid (condensed) phase and a positive isotope effect exists. When $|\Delta H_D| - |\Delta H_H| < 0$, deuterium is concentrated in the gas phase and a negative isotope effect occurs. As is evident from Tables 2.2, 3, for the majority of systems reversibly sorbing hydrogen one observes a positive isotope effect, i. e. $P_{T_2} < P_{D_2} < P_{H_2}$ at given hydrogen content.

It should be noted that in a wide temperature range a number of systems can display inversion of the isotope effect. This question will be considered more fully in the next chapter.

It is assumed that finding the thermodynamic values of the Me(IMC)–H$_2$ system from relationship (2.6) is the classic experimental method. The only assumption used in this method, namely that ΔH and ΔS do not depend on temperature, is valid for a limited number of compounds, and in some cases can lead to essential errors in the determination of ΔH.

As opposed to the P-C-T method, the equilibrium calorimetric method (ECM) is more widely applicable.

2.3 Equilibrium Calorimetric Method
for Studying Hydrogen–Me(IMC) Systems

ECM was devised and originally applied to study the thermodynamics of Me–H_2 systems by *Kleppa* and co-workers in 1973–1977 [2.42–45] by combining the equilibrium (P-C-T) method and the calorimetric method in a single experimental setup. Thereafter it has been used by a number of authors [2.46–49] for studying the thermodynamics of IMC–H_2 systems.

In contrast to the P-C-T method, ECM gives a possibility to define the parameters ΔH, ΔS, ΔG of the hydride-formation reaction in a single experiment, significantly improves the accuracy and reliability of the determination of ΔH and allows one to establish the dependence of ΔH on the hydrogenation degree n.

These advantages enable one to use ECM in studies of surface degradation, hysteresis, and other phenomena accompanying the hydride-formation process. In [2.50–53] by means of ECM both the systems ZrB_2–H_2 (where B is a transition metal: V, Cr, Mn) and pseudo-binary and hyper-stoichiometric compounds based on these systems were investigated. Variation of the thermodynamic and kinetic parameters was performed by substituting different components of the alloy.

These systems offer several advantages: high rates of hydrogen sorption–desorption, significant capacity (200–250 cm^3 H_2/g IMC), and considerable values of separation factors.

The samples were prepared by the method of electric arc melting in an atmosphere of pure argon. A total of 9 compounds of composition ZrB_2 were synthesized. In the process of the initial charge preparation, the components were mixed in the stoichiometric proportion with the exception of compounds containing Mn. Taking into account its high volatility manganese was added with an excess in the order of 5–8 wt.%. The correspondence between IMC and their stoichiometric composition was checked by weighing the sample before and after melting.

To homogenize the sample IMC were held at $T = 1320\,\mathrm{K}$ for 48 hours.

The data of the X-ray phase analysis of the samples are presented in Table 2.4 in comparison with the data of other authors. Reasonable agreement can be noted in the values of the parameters of the crystal lattices of IMC.

The composition of IMC has an effect not only on the absolute values of equilibrium pressures of hydrogen but also on the form of the dependencies of P_{H_2} and ΔH_H on n, where n is the number of H-atoms per single atom of IMC.

To establish general thermodynamic relationships for the interaction of compounds of the AB_2 type the samples of binary compounds ZrV_2, $ZrCr_2$, $ZrMn_2$ were investigated. For these compounds both the concentration dependencies of the heat of formation and decomposition of hydrides in the temperature range from 333 to 433 K (Fig. 2.2) and the isotherms of sorption-desorption (Fig. 2.3) were obtained.

Table 2.4. IMC structure and lattice parameters

IMC	Structure	Experimental data		Literature data		Reference
		$a_0/Å$	$c_0/Å$	$a_0/Å$	$c_0/Å$	
ZrV_2	C15	7.440	–	7.438	–	[2.54]
$ZrV_{0.5}Cr_{1.5}$	C15	7.257	–	–	–	–
$ZrV_{0.2}Cr_{1.8}$	C15	7.216	–	–	–	–
$ZrCr_2$ (baked)	C15	7.198	–	7.19	–	[2.54]
$ZrCr_2$ (unbaked)	C14	5.092	8.252	5.107	8.272	[2.55]
ZrVMn	C14	5.080	8.318	–	–	–
ZrMnCr	C14	5.042	8.281	–	–	–
$ZrMn_2$	C14	5.016	8.222	5.030	8.370	[2.56, 57]
$Zr_{0.8}Mn_{0.2}Mn_2$ (unbaked)	C14	4.989	8.199	–	–	–
$Zr_{0.8}Mn_{0.2}Mn_2$ (baked)	C14	4.968	8.149	5.004	8.208	[2.56, 57]
$Zr_{0.7}Mn_{0.3}Mn_2$	C14	4.950	8.120	–	–	–

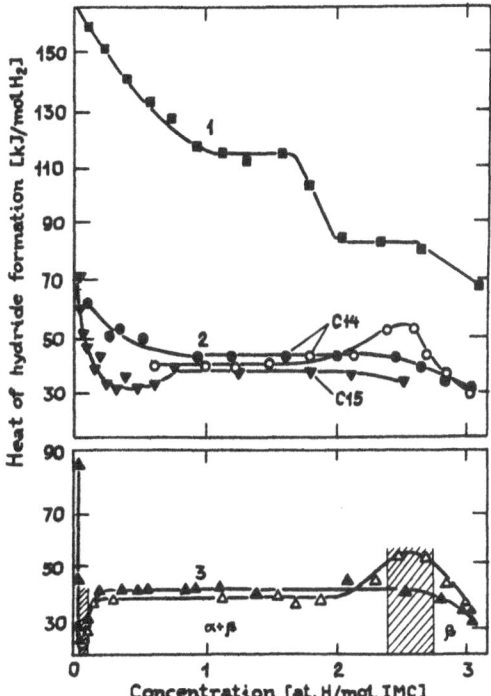

Fig. 2.2. Concentration dependencies of enthalpy of formation and decomposition of IMC hydrides of the ZrB_2 type: (1) ZrV_2 (433 K), (2) $ZrCr_2$ (433 K), (3) $ZrMn_2$ (333 K)

In Fig. 2.2 values of ΔH_H were determined for the middle of the concentration range, which corresponds to each step on the isotherm in Fig. 2.3 (i.e., $n = \sum_{i=1}^{N} \Delta n_{i-1} + \Delta n_i / 2$, where $i = 1, 2 \dots N$ is the number of the step, Δn_i is the size of each step).

For the initial portion of the isotherms for $ZrCr_2$ and $ZrMn_2$ and for the system ZrV_2–H_2 no desorption data are available concerning the duration of desorption at low P_{H_2}.

Individual properties of IMC are reflected in differences in the forms of the dependencies $\Delta H_H = f(n)$ and $P_{H_2} = f(n)$, which exhibit themselves in the following manner:

1. The presence of structure allotropy can be observed only for $ZrCr_2$, which exists both as FCC(C15) formed after high-temperature bake-out and as HCP(C14) structures. This results in some distinctions in the form of ΔH_H-curves in Fig. 2.2 for this compound and in somewhat higher absolute values of ΔH_H for the C14-structure than for C15 (44 and 38 kJ/mol H_2, respectively). ZrV_2 has only C15-structure and $ZrMn_2$ has only C14-structure.

2. Only ZrV_2 possesses the ability to fill different interstices, where hydrogen sequentially fills in interstices of composition Zr_2V_2 and ZrV_3. $ZrCr_2$ and $ZrMn_2$ are typified by filling interstices of composition Zr_2Cr_2 and Zr_2Mn_2 [2.58]. That is why on the curve ΔH_H vs. n (Fig. 2.2) ZrV_2 exhibits two characteristic plateaus and $ZrMn_2$ exhibits only one.

3. The ability to undergo isomorphous transformations when interacting with hydrogen is most clearly defined for $ZrMn_2$, where separation into α- and β-phases takes place. This in turn manifests itself as significant hysteresis of pressures of sorption and desorption. In $ZrCr_2$ one observes a similar tendency to phase separation, and the system ZrV_2–H_2 is a solid α-solution of hydrogen in ZrV_2 [2.55].

The behaviour of the system $ZrMn_2$–H_2 is similar to the majority of the IMC–H_2 systems having α- and β-phases and, hence, it is convenient to use it to determine the regions of phase borders from the characteristic bends in the dependence $P_{H_2} = f(n)$. Thus one can divide the concentration dependence of ΔH_H (Fig. 2.2) into several sections and analyze each of them individually.

1. $|\Delta H_H|$ decreases abruptly until it reaches a constant value ($0 < n < 0.06$). This characteristic of ΔH_H differs from the Me–H_2 systems, for which the initial portion of the isotherms is concerned with the formation of α-soltuion and $|\Delta H_H|$ increases with increasing n [2.42–45]. It is to be noted that a decreasing $|\Delta H_H|$ is typical for all binary compounds, although the concentration ranges in which this effect is observed can differ widely. The experimental data allow one to suppose that the dissolving of hydrogen in the IMC studied is followed by a process whose heat effect is considerably higher.

2. In the range of concentrations $0.06 < n < 0.1$, which corresponds to the $\alpha/(\alpha+\beta)$ border one observes an increase of $|\Delta H_H|$ with increasing n from ΔH_H^{α}, characterizing the formation of an α-solution of hydrogen in $ZrMn_2$, to $\Delta H_H^{\alpha-\beta}$

showing that in this system structure transition occurs. Such a transition is absent in ZrV_2 and therefore the dependence ΔH_H on n has no corresponding bend in the relevant portion and for $ZrCr_2$(C15) one observes a minor difference between ΔH_H^α and $\Delta H_H^{\alpha-\beta}$ of the order of 5 kJ/mol H_2 and an extended minimum on the plot $\Delta H_H = f(n)$.

3. In the range corresponding to the α–β transition ($0.1 < n < 2.44$) ΔH_H remains practically constant forming a plateau. Such a plateau is also observed in the $ZrCr_2$–H_2 system.

4. Essential differences in the values of ΔH_H of sorption and desorption of the order of 10–12 kJ/mol H_2 are grouped on the border $(\alpha + \beta)/\beta$ ($2.44 < n < 2.73$) and a similar situation also occurs in the $ZrCr_2$–H_2 system.

5. Solution of hydrogen in the β-phase ($n > 2.73$) is followed by some decrease of ΔH_H with increasing n.

Thus the interaction of hydrogen with IMC is related to a number of phenomena effecting the thermodynamics of the process, which in turn results in changes of the heat effects of the reaction.

Based on an analysis of the dependence of ΔH_H on n for the binary compounds it is possible to reveal the common features characteristic for the interaction of IMC with hydrogen. These are (a) a sharp drop of $|\Delta H_H|$ on the initial portion of the concentration dependence reflecting the segregation and destruction processes in IMC, which distinguish them from the Me H_2 systems; (b) differences in ΔH_H of sorption and desorption on the phase borders concerned with the hysteresis also characteristic for the Me-H_2 systems.

ECM enables one to estimate these phenomena quantitatively.

The decrease of $|\Delta H_H|$ with increasing concentration on the initial portion is discussed in [2.47, 48, 59] using the "capture" model; higher values of ΔH_H in comparison to the α-phase are assumed to be related to the formation of "traps", i.e., interstices enriched with hydride-forming component. It is assumed that the decrease of the heat released in the process of IMC interaction with hydrogen is caused by progressive filling of the traps.

Two types of interstices are considered: normal interstices of $LaNi_2$ and trap interstices. It is assumed that only the trap type takes part in the interaction.

Thus the experimental value of ΔH_H at any n can be presented as

$$\Delta H_H = \Delta H_H^t f_H^t + \Delta H_H^f f_H^f, \tag{2.26}$$

where f_H^t and f_H^f are the fractions of hydrogen atoms entering the traps and normal interstices, respectively, at a given n, and ΔH_H^t and ΔH_H^f are the changes of enthalpy concerned with the capture of hydrogen by the traps (they can be found by extrapolation of the experimental dependence $\Delta H_H = f(n)$ to intersect with the ordinate) and by normal interstices of $LaNi_2$.

To process the data the authors [2.47, 48, 59] modified the model proposed in [2.60] and made a number of assumptions: the H–H interaction is neglected, blocking effects are assumed absent and S_H^0 is taken to be equal for traps and normal interstices, where S_H^0 is the excess entropy at $n \to 0$.

It is to be noted that good agreement between the calculated and the experimental data is found both for the dependence of ΔH_H on n and also for ΔS_H as a function of n.

The fraction of trap interstices was determined as $\beta_t = 0.15$ per mole of LaNi$_2$. This model has also been applied by other authors [2.49, 61].

The drawback of the "capture" model as applied to specific systems is related to the unexplored nature and type of the traps being filled, and manifests itself in the determination of ΔH_H^t by the extrapolation of the experimental dependence of ΔH_H on n to intersect with the ordinate. It is obvious from Fig. 2.3 that $n \to 0$ this dependence nearly asymptotically approaches to the y-axis, which can result in errors not only in the value of ΔH_H^t but also in determination of the fraction of captured hydrogen atoms (since the energy parameter $\varepsilon = \Delta H_H^t - \Delta H_H^f$ is used for this purpose).

In systems such as $Zr(V_y Cr_{1-y})_2$–H$_2$, where the capture of hydrogen is observed over a wide range of concentrations, these errors can be very considerable.

As opposed to the "capture" model, which does not consider the causes of segregation in IMC, let us assume that it is merely interaction with hydrogen that can result in partial destruction of IMC and that segregation is the first step of this process and can characterize the insensitivity of IMC to the hydrogen effect. The

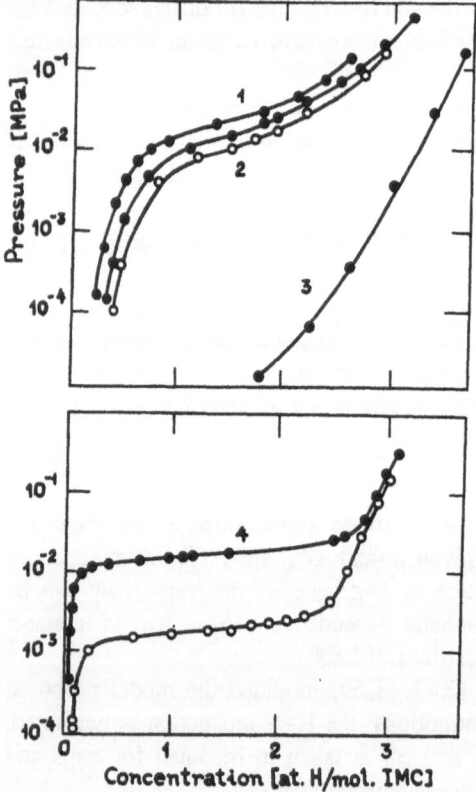

Fig. 2.3. Isotherms of hydrogen sorption–desorption on IMC of the ZrB$_2$ type: (*1*) ZrCr$_2$(C15), (*2*) ZrCr$_2$(C14), (*3*) ZrV$_2$, (*4*) ZrMn$_2$; • sorption, ○ desorption

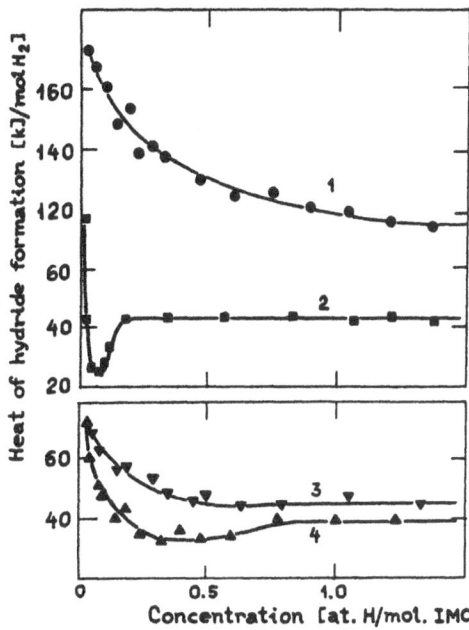

Fig. 2.4. Initial portions of dependencies of enthalpy of ZrB_2 hydrides formation on hydrogen concentration: (1) ZrV_2 (433 K), (2) $ZrMn_2$ (333 K), (3) $ZrCr_2$(C14), (4) $ZrCr_2$(C15) at 433 K

experiments carried out on laser desorption in hydrogen atmosphere corroborate the degradation of the surface layer of ZrB_2 in hydrogen.

Figure 2.4 presents the spectra of the ZrV_2 sample after holding for 2 h in hydrogen at 170 °C. Comparison of the spectra shows that in the main phase $ZrV_2(a)$, $Zr(b)$ and $V(b)$ are the degradation products existing in the surface layer.

Since during the experiments formation of ZrH_2 for all IMC investigated and VH_n for vanadium-containing compounds is possible, one should estimate the fraction of hydrogen reacted with vanadium from the initial portion of the isotherms. In line with the data on phase equilibrium in the $V–H_2$ system [2.20], the formation of hydrides of composition $VH_{0.1}$ for ZrV_2, $VH_{0.5}$ for $ZrV_{0.5}Cr_{1.5}$, and $VH_{0.6}$ for $ZrV_{0.2}Cr_{1.8}$ becomes possible. Taking into account $\Delta H_H = 67$ kJ/mol H_2 [2.22] the possible contribution to the heat effect at the cost of VH_n formation may reach 6 % depending on the IMC composition, which is neglected in the following calculations.

The approach proposed can be represented as a limiting case of the "capture" model, where interstices of Zr act as the traps

$$\Delta H_H = \Delta H_{Zr-H} f_H^{Zr} + \Delta H_{ZrB_2-H} f_H^{ZrB_2} \tag{2.27}$$

$$f_H^{Zr} + f_H^{ZrB_2} = 1 \tag{2.28}$$

where f_H^{Zr} and $f_H^{ZrB_2}$ are the fractions of hydrogen entering Zr and ZrB_2 interstices, respectively, and ΔH_{Zr-H} and ΔH_{ZrB_2-H} are the heats of hydride formation for Zr (determined calorimetrically and equal to -225 kJ/mol H_2) and for the corresponding IMC.

The experimental data analysis makes it possible not only to define the fraction and amount of hydrogen captured but also to estimate the fraction released as a result of Zr destruction in the systems explored starting from the highest degree of hydrogenation $n = 2$ for Zr.

The results of the experimental data analysis using (2.27, 28) are summarized in Table 2.5.

Table 2.6 shows that the "capture" model predicts significantly greater capture of hydrogen than in the case of hydride formation.

Table 2.5. Experimental and computed parameters for studied IMC

IMC	Lattice type	T K	$\|\Delta H_H^\alpha\|$ kJ/mol H$_2$	$\|\Delta H_H^{\alpha-\beta}\|$ kJ/mol H$_2$	η_H^{Zr} $n = 1.5$ %	β^{Zr} %
ZrV$_2$	FCC(15)	433	–	115.0	12.8	9.2
ZrV$_{0.2}$Cr$_{1.8}$	(C15)	433	–	57.0	7.0	5.2
ZrV$_{0.5}$Cr$_{1.5}$	(C15)	433	–	48.0	3.4	2.6
ZrCr$_2$	(C15)	433	33.0	38.0	1.6	1.2
ZrVMn	hex.(C14)	433	55.0	70.0	1.6	1.2
ZrCrMn	(C14)	333	27.2	43.0	0.8	0.6
ZrMn$_2$	(C14)	333	25.0	42.5	0.9	0.7
Zr$_{0.8}$Mn$_{0.2}$Mn$_2$	(C14)	333	22.7	29.5	0.07	0.05
Zr$_{0.7}$Mn$_{0.3}$Mn$_2$	(C14)	293	19.1	27.5	0.05	0.04
ZrCr$_2$	(C14)	333	–	48.0	2.6	2.0

η_H^{Zr} is the ratio between the amount of hydrogen captured by zirconium and the overall hydrogen concentration, n is assumed equal for all compounds ($n = 1.5$), β^{Zr} is the amount of Zr forming hydride per overall amount of Zr in IMC

Table 2.6. Relative fractions of "captured" hydrogen in Zr(V$_y$Cr$_{1-y}$)$_2$ calculated by different methods

IMC	η_H (%) at $n = 1.5$		
	"capture" model	hydride formation	isotope exchange
ZrCr$_2$(C15)	7.4	1.6	2.0
ZrV$_{0.2}$Cr$_{1.8}$	10.8	3.4	3.8
ZrV$_{0.5}$Cr$_{1.5}$	18.2	7.0	9.0
ZrV$_2$	23.4	12.8	14.1

To corroborate the quantitative evaluations obtained using two models, isotope exchange in Zr(V$_y$Cr$_{1-y}$)$_2$–HT systems is investigated, because the effect of "capture" most clearly manifests itself for exactly these IMC, as seen from Table 2.5.

Isotope exchange in a loop reactor enables one to determine the amount of hydrogen bound at normal interstices and at Zr interstices since, in the experiment at $T = 433$ K, the rate of isotope exchange in the ZrH_2–HT system is very small [2.62], i.e., protium captured by Zr interstices practically does not exchange with tritium of the gas phase, whereas at such temperature the separation factor is close to 1 for the main phase of $Zr(V_yCr_{1-y})_2$ [2.63].

N_S^{in} is determined by using the mass balance equation:

$$N_G^C I_G^0 = \left(N_G^C + N_G^R\right) I_G^\infty + N_S^{in} I_S^{in} \tag{2.29}$$

from which, considering $I_S^\infty = I_G^\infty$ (since $\alpha = 1$), we derive:

$$N_S^{in} = \left(N_G^C + N_G^R\right) - \frac{N_G^C I_G^0}{I_G^\infty}, \tag{2.30}$$

where N_G^C, N_G^R are the number of moles of hydrogen in the gas phase in the cycle and in the reactor (at 433 K); I_G^0, I_G^∞, I_S^∞ are initial and equilibrium activity in gas and solid phases, (counts/sec); N_S^{in} is the number of moles of hydrogen in the solid phase taking part in exchange and bound at normal interstices of $Zr(V_yCr_{1-y})_2$.

The amount of hydrogen not involved in exchange N_S^{notin} is defined as

$$N_S^{notin} = N_S^0 - N_S^{in}, \tag{2.31}$$

where N_S^0 is the overall amount of hydrogen (mol) in the solid phase defined by the volume method.

The fraction of hydrogen in the solid phase not included in exchange, reduced to $n = 1.5$ for each IMC, is defined as

$$\eta_H^{Zr} = \frac{N_S^{notin}}{N_S^0} \quad \text{or generally} \quad \eta_H^t = \frac{N_S^{notin}}{N_S^0}. \tag{2.32}$$

The values N_S^{notin} and N_S^{in} can be found by the mass balance equation and the gas phase activity after desorption at high temperatures, i.e., when Zr hydride is completely destroyed.

The two methods of $\eta_H^{Zr}(\eta_H^t)$ determination are found to agree adequately for all four systems ($\pm 10\%$ of η_H). The values obtained are averaged and summarized in Table 2.6 as compared with η_H^{Zr} and η_H^t obtained from the calorimetric measurements.

Good coincidence of η_H in the two last columns of Table 2.6 and the essential difference from η_H^t determined with the "capture" model enables us to conclude that partial destruction takes place when compounds of the ZrB_2 type interact with hydrogen to form zirconium hydride.

In the calculation of the degree of destruction (2.27, 28) presented in Tables 2.5, 6, values of ΔH_{ZrB_2} were ignored and it was assumed that this parameter does not essentially affect β^{Zr}, which is mainly determined by the width of the concentration range where decrease of $|\Delta H_H|$ occurs.

Consideration of ΔH_{ZrB_2} for any IMC has to result in an increase of β^{Zr}, since decomposition of IMC is a process accompanied by heat absorption and, hence, the value $|\Delta H_{Zr-H}|$ will decrease and f_H^{Zr} will increase. To demonstrate the importance of this term, values of β^{Zr}, considering IMC heat of formation, must be compared with the data in Table 2.5

Table 2.7 demonstrates the data calculated in [2.64, 65] and obtained experimentally by the Knudsen effusion method together with mass-spectrometric analysis of the gas phase (high-temperature mass-spectrometry) for the two binary compounds ZrCr$_2$ and ZrMn$_2$ [2.66].

Table 2.7. Heat effects of IMC of AB$_2$ type formation

IMC	ΔH_{ZrB_2} (kJ/mol)		
	Calculation by Miedema [2.64]	Calculation by Shaltiel [2.65]	Effusion method
ZrV$_2$	−20.0	−4.6	−
ZrCr$_2$	−41.8	−16.3	−21.0
ZrMn$_2$	−57.6	−21.3	−36.0

Though the error in the effusion method is rather large (± 10 kJ/mol), it should be noted that, for the given IMC, values of heats of formation were obtained for the first time. This enables one to compare them with those calculated by Miedema and Shaltiel and they are found to give better agreement with the data of the latter.

Consideration of ΔH_{ZrB_2} in (2.27) results in some increase in the degree of destruction from 1.2 % to 1.4 % for ZrCr$_2$ and from 0.7 % to 0.8 % for ZrMn$_2$, i.e., this number can be ignored in quantitative evaluations.

Figure 2.5 demonstrates the destruction degree for β^{Zr} and the heat of hydride-formation $|\Delta H_H|$ as a function of the number of valence electrons per metal atom of the IMC.

The dependencies presented have parallel character; in this case three characteristic regions can be resolved. The first is described by an abrupt decrease of $|\Delta H_H|$ and β^{Zr} in the range of e/m from 4.67 to 5.33 (from ZrV$_2$ to ZrCr$_2$, from 115 to 38 kJ/mol H$_2$ and from 9.2 to 1.2 %, respectively), i.e. the range of instability manifests itself. The second range ($5.33 < e/m < 6.00$) is typified by some decrease of $|\Delta H_H|$ and β^{Zr} (transition region) and finally in the third range ($e/m > 6.00$) β^{Zr} tends to zero with further decrease of $|\Delta H_H|$ (region of stability).

However, it is to be noted that a calculation of the ratio e/m for hyper-stoichiometric IMC must take into account substitution of hydride-forming Zr by the transition metal Mn in the crystal lattice of AB$_2$. In this case the number of valence electrons of Mn taking part in the formation of the metal bond instead of Zr can be equal to the number of valence electrons of Zr, i.e. four instead of seven. With this approach, e/m for all stoichiometric compounds of the Zr$_y$Mn$_{1-y}$Mn$_2$ row is constant and equal to 6.

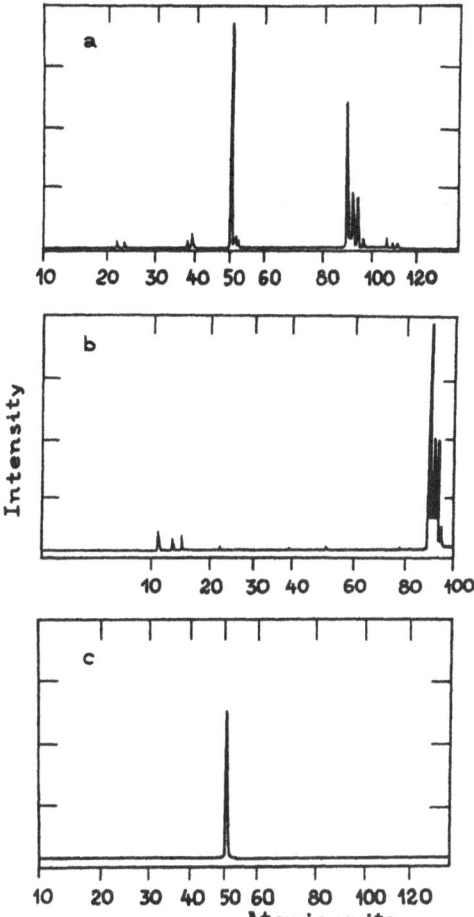

Fig. 2.5a–c. Spectra of laser desorption of the superficial layer after exposure to hydrogen: (a) main phase of ZrV_2, (b) Zr, (c) V

The values of β^{Zr} presented in Fig. 2.5 change weakly at $e/m > 6$; at the same time $|\Delta H_H|$ decreases from 42.5 to 27.5 kJ/mol H_2. A similar pattern is observed for the isoelectronic compounds $ZrCr_2$ and $ZrVMn$: at $e/m = 5.33$ ΔH_H differs essentially (up to 30 kJ/mol H_2), whereas β^{Zr} is equal to 1.2 % for both. This enables one to assume that the electronic factor is decisive for IMC of the ZrB_2 type which are resistant to the hydrogen effect.

It should be noted that the linearity of the dependence $|\Delta H_H|$ on e/m on different portions corresponding to pseudo-binary compounds also remains in Fig. 2.6, where y is the fraction of one of the components B in the pseudo-binary compounds described by the formula $Zr(B'_y B''_{1-y})_2$. Obviously, this can be extended to other pseudo-binary compounds and enables one to determine the heat effects of hydride-formation for any similar compound knowing only ΔH_H for the binary alloys AB'_2 and AB''_2 at a definite temperature.

The data presented in Table 2.5 and in Fig. 2.5 show that the resistance of the binary compounds to the hydrogen effect increases in the series ZrV_2–$ZrCr_2$–$ZrMn_2$.

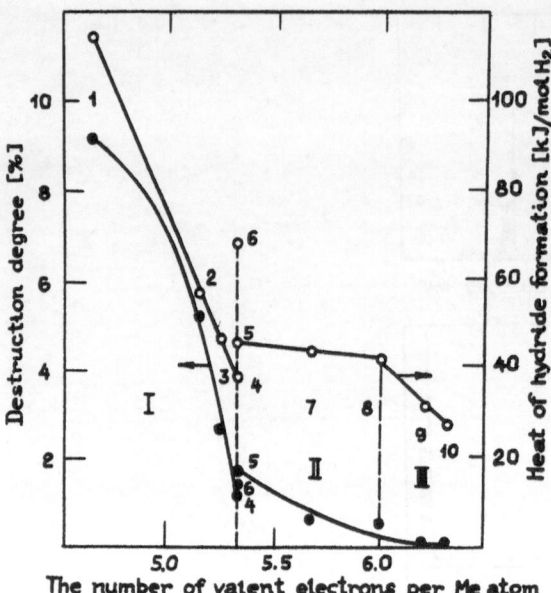

Fig. 2.6. Interrelation between the heat of formation, destruction degree, and amount of valent electrons per metal atom in IMC of the ZrB_2 type: (1) ZrV_2, (2) $ZrV_{0.5}Cr_{1.5}$, (3) $ZrV_{0.2}Cr_{1.8}$, (4) $ZrCr_2$(C15) (all the data relate to $T = 433$ K), (5) $ZrCr_2$(C14) (333 K), (6) $ZrVMn$ (433 K), (7) $ZrCrMn$, (8) $ZrMn_2$, (9) $Zr_{0.8}Mn_{0.2}Mn_2$ (all the data relate to $T = 333$ K), (10) $Zr_{0.7}Mn_{0.3}Mn_2$ (293 K)

The behaviour of the thermodynamic parameters in the pseudo-binary compounds $Zr(V_yCr_{1-y})_2$, $Zr(Cr_yMn_{1-y})_2$, $Zr(V_yMn_{1-y})_2$ in Fig. 2.5 and in Tables 2.5, 6 corroborates this conclusion and injection of hyper-stoichiometric Mn practically totally suppresses the destruction.

The above conclusions were corroborated by means of the laser desorption method [2.67]; the data are summarized in Tables 2.8, 9.

Table 2.8. Composition of non-activated IMC in the bulk and on the surface

IMC	Surface	Δx_{Zr} (%)	Bulk	Δx_{Zr} (%)
ZrV_2	$Zr_{2.17}V_{0.83}$	+117	$Zr_{1.20}V_{1.80}$	+20
$ZrCr_2$	$Zr_{1.40}Cr_{1.60}$	+40	$ZrCr_2$	±0
$ZrMn_2$	$Zr_{1.80}Mn_{1.20}$	+80	$Zr_{1.14}Mn_{1.86}$	+14
$Zr_{0.8}Mn_{2.2}$	$Zr_{2.11}Mn_{0.89}$	+164	$Zr_{0.80}Mn_{2.20}$	±0
$Zr_{0.7}Mn_{2.3}$	$Zr_{1.47}Mn_{1.53}$	+110	$Zr_{0.65}Mn_{2.35}$	−7
$ZrCrMn$	$Zr_{1.20}Cr_{1.00}Mn_{0.80}$	+20	$Zr_{0.83}Cr_{1.34}Mn_{0.83}$	−17
$ZrVMn$	$Zr_{1.40}V_{0.80}Mn_{0.80}$	+40	$ZrVMn$	±0
$ZrV_{0.2}Cr_{1.8}$	$Zr_{1.32}V_{0.24}Cr_{1.45}$	+31	$ZrV_{0.2}Cr_{1.8}$	±0
$ZrV_{0.5}Cr_{1.5}$	$Zr_{1.10}V_{0.55}Cr_{1.35}$	+10	$ZrV_{0.5}Cr_{1.5}$	±0

Table 2.9. Composition of activated IMC in the bulk and on the surface

IMC	Surface	Δx_{Zr} (%)	Bulk	Δx_{Zr} (%)
ZrV_2	$Zr_{2.23}V_{0.77}$	+123	$Zr_{1.65}V_{1.35}$	+65
$ZrCr_2$	$Zr_{1.60}Cr_{1.40}$	+60	$Zr_{1.60}Cr_{1.40}$	+60
$ZrMn_2$	–	–	–	–
$Zr_{0.8}Mn_{2.2}$	$Zr_{2.11}Mn_{0.89}$	+164	$Zr_{0.98}Mn_{2.02}$	+23
$Zr_{0.7}Mn_{2.3}$	$Zr_{1.65}Mn_{1.35}$	+136	$Zr_{1.14}Mn_{1.86}$	+63
$ZrCrMn$	–	–	–	–
$ZrVMn$	$Zr_{1.30}V_{0.80}Mn_{0.90}$	+30	$Zr_{1.26}V_{0.97}Mn_{0.77}$	+26
$ZrV_{0.2}Cr_{1.8}$	$Zr_{1.22}V_{0.31}Cr_{1.47}$	+22	$Zr_{1.18}V_{0.40}Cr_{1.42}$	+18
$ZrV_{0.5}Cr_{1.5}$	$Zr_{1.08}V_{0.59}Cr_{1.33}$	+8	$Zr_{1.18}V_{0.49}Cr_{1.33}$	+18

As can be seen from Figs. 2.7, 8, the superficial layers (thickness $\leq 0.1\,\mu m$) of both activated and non-activated samples show enhanced Zr contents (Δx_{Zr}) compared with the stoichiometry. This enhancement is observed to be higher with oxygen than with hydrogen impurities on the surface.

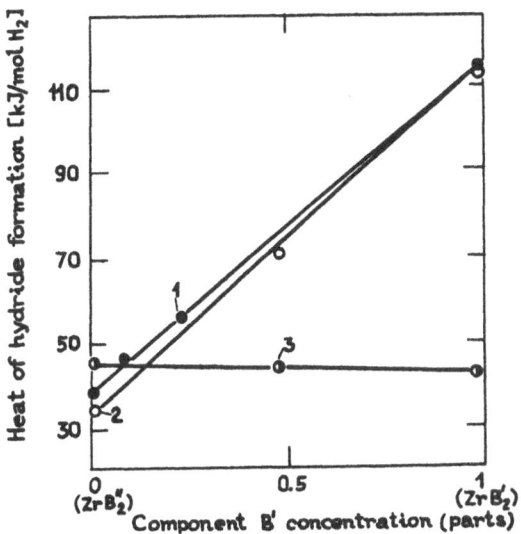

Fig. 2.7. Influence of partial B-component substitution on the heat of hydride formation: (*1*) $Zr(V_yCr_{1-y})_2$, (*2*) $Zr(V_yMn_{1-y})_2$, (*3*) $Zr(Cr_{1-y}Mn_y)_2$ at $T = 333\,K$

At a depth of $< 0.3\,\mu m$ the deviation from the stoichiometric composition is significantly higher for non-activated samples than for activated ones. This on the one hand, testified that partial destruction of IMC takes place, qualitatively corroborating the conclusions presented above, and, on the other hand, points to segregation processes in the bulk of IMC caused by the hydrogen effect.

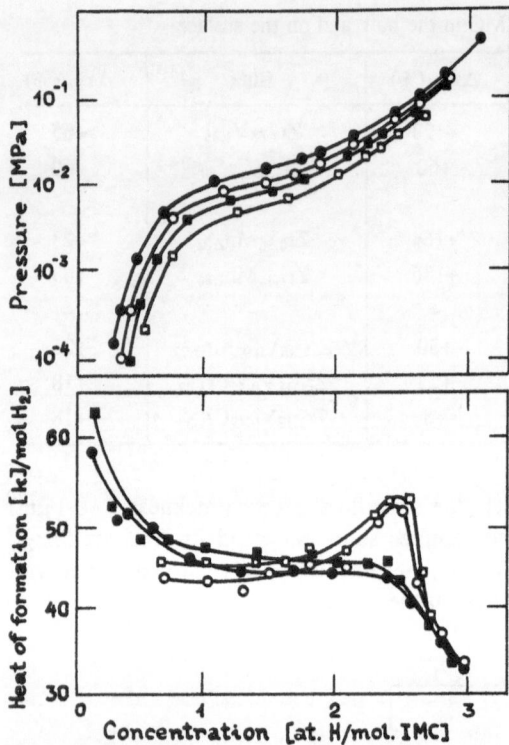

Fig. 2.8. Isotherms of sorption and dependencies of heat of formation on the hydrogen concentration for $ZrCr_2–H_2$ and $ZrCr_2–D_2$ systems at 300 K ($o, \bullet$ $ZrCr_2–H_2$; $\square, \blacksquare$ $ZrCr_2–D_2$; empty symbols relate to desorption, filled ones relate to sorption)

From the calorimetric investigations it follows that the greatest destruction degree is expected for ZrV_2 and the lowest one for hyper-stoichiometric compounds $Zr_{0.8}Mn_{0.2}Mn_2$ and $Zr_{0.7}Mn_{0.3}Mn_2$. This is corroborated by the data presented in Tables 2.8, 9.

Segregation in the bulk of IMC is typical for these compounds and for others, but, as opposed to Cr and V, manganese has the property of replacing Zr in the crystal lattice of IMC (synthesis of compounds of the $Zr_{1-y}Mn_yMn_2$ series is founded on exactly this fact). So, segregation in the bulk of these IMC results not in the appearance of pure Zr and Mn-phases, as one may conclude from the behaviour of ZrV_2, but in the formation of solid solutions of common composition $Zr_{1-y}Mn_yMn_2$, where $y \leq 0$ in the vicinity of the surface, and at depths of $> 0.3\,\mu m$ it corresponds to its stoichiometric value.

Such an approach accounts for weak capture of hydrogen by zirconium in these compounds and also explains the slope of the plateau on the plot $\Delta H_H = f(n)$ caused by filling the interstices with different Mn content.

The method of equilibrium calorimetry enables one to estimate the value of the isotope effect of hydrogen in studies of protium and deuterium sorption. The work

[2.24] was among the first to deal with the study of the Ti–$H_2(D_2)$ systems by means of high-temperature calorimetry; however, at high temperatures the values of isotope effects are small, leading to significant errors in their determination.

Let us illustrate the possibilities of ECM by the example of the $ZrCr_2$–$H_2(D_2)$ system [2.50]. Figure 2.8 presents the dependencies $\lg P_{H_2(D_2)} = f(n)$ and $|\Delta H_{H(D)}| = f(n)$. It is evident that these dependencies have similar character and that within practically the total range $|\Delta H_D| > |\Delta H_H|$ by an amount 1.5–2 kJ/mol H_2. It is known that isotope effects are most strongly manifested at low temperatures. However, in this case, an abrupt decrease of the rate of hydride formation takes place and the accuracy of $P_{H_2(D_2)}$ and $\Delta H_{H(D)}$ determination falls markedly. For this reason a temperature of 300 K is chosen for carrying out the experiments.

Since the difference between the heats of hydride- and deuteride-formation is small, namely, of the order of 3 % of ΔH_H, saturation by protium and deuterium is performed in one step up to the capacity of $n = 1.5$ for more precise calculation of the thermodynamic parameters. The values obtained for the enthalpy are the following: $\Delta H_H = -45.8$ kJ/mol H_2, $\Delta H_D = -47.3$ kJ/mol D_2 and the values of entropy calculated by (2.6) are $\Delta S_H = -109.0$ J/mol K, $\Delta S_D = -111.1$ J/mol K. These values together with (2.13) allow one to estimate the value of the isotope effect for this system $P_{D_2}^{1/2}/P_{H_2}^{1/2} = 1.1$, which adequately agrees with the data obtained by the method of single adjusting for equilibrium (Chap. 3).

3. Isotope Equilibrium of Hydrogen with Hydride Phases (Thermodynamic Isotope Effect)

3.1 Thermodynamic Isotope Effect and Its Connection with Phase Equilibrium

As described above, the formation of the hydride phase of a metal or IMC is escorted by dissociation of hydrogen molecules on the surface and implantation of atoms in interstices of the crystal lattice. In crystal lattices of metals and IMC hydrogen atoms can occupy both tetrahedral and octahedral interstitial positions. As a rule, tetrahedral interstices are filled at low temperatures in IMC at any hydrogen concentration (up to values corresponding to maximal d-zone filling). At high temperatures, hydrogen fills primarily octahedral positions.

By considering potential wells and the zero-point energies of gaseous hydrogen molecules and atoms in tetrahedral and octahedral interstices, the cause of the thermodynamic isotope effect becomes clear. Let us consider the systems H_2–$TiMn_{1.5}$ (hydrogen atoms occupying tetrahedral positions) and H_2–Pd (hydrogen atoms located in octahedral interstices) as an example. In the first case a "positive" isotope effect is observed, in which the heavy isotope prefers the solid phase. In the second case the heavy isotope is concentrated in the gas phase, and this is referred to as the "negative" isotope effect. Figure 3.1 qualitatively shows the form of potential wells for molecular hydrogen and hydrogen atoms in hydride phases of the IMC $TiMn_{1.5}$ and the metal Pd. For hydride phases values of the zero-point energy are obtained from inelastic neutron scattering data in terms of the three-fold degenerate mode of hydrogen atoms in the crystal lattice (for $TiMn_{1.5}H_{2.5}$, $\hbar\omega_H = 145.8$ and $\hbar\omega_D = 103.4$ meV [3.1] and for $PdH_{0.65}$, $\hbar\omega_D = 58$ and $\hbar\omega_D = 40$ meV [3.2]).

As is evident from Fig. 3.1, for isotope substitution of protium atoms with deuterium ones the change of zero-point energy in the hydride phase of the IMC is more considerable than in gaseous hydrogen. That is why deuterium atoms prefer to locate in the hydride phase and protium atoms in the gas. For implantation of hydrogen atoms in octahedral interstices of the crystal lattice of Pd, the reverse situation is observed. Namely, substitution of protium with deuterium results in a smaller zero-point energy change in the hydride phase in comparison with gaseous hydrogen and hence deuterium is enriched in the gas phase. At high deuterium concentrations the change of zero-point energy for the isotope substitution H–D in the hydride phase has to compare with the difference of zero-point energies of molecules HD and D_2, which appears to be higher ($\Delta\varepsilon_0 = 43$ meV) than for the

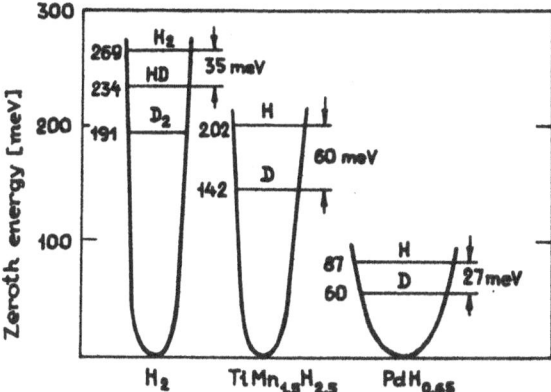

Fig. 3.1. Zero-point energies of isotope modifications of gaseous hydrogen molecules and hydrogen atoms in tetrahedral interstitial sites of $TiMn_{1.5}H_{2.5}$ and in octahedral interstitial sites of $PdH_{0.65}$

molecules H_2 and HD ($\Delta\varepsilon_0 = 35\,meV$). It leads to an increase of the thermodynamic effect in the H_2–Pd system as the deuterium concentration increases. For the H_2–$TiMn_{1.5}$ system the situation is reverse. The effects considered are more clearly defined for the isotope exchange H–T.

The separation factor for a binary mixture gives a quantitative measure of the isotope effect. At thermodynamic equilibrium it manifests itself in the ratio of relative concentrations of the isotopes being distributed between two phases in equilibrium:

$$\alpha = \left(\frac{x}{1-x}\right) \Big/ \left(\frac{y}{1-y}\right), \tag{3.1}$$

where x, y are the atomic fractions of the heavy hydrogen isotope (deuterium or tritium) in the hydride and gas phase, respectively. In most cases the solid phase is enriched with the heavy isotope. The exception is some IMC and, as a rule, metals.

The thermal isotope effect in hydrogen–hydride-phase systems is related to the difference of isotherms of hydrogen sorption by the solid phase. As opposed to the most famous case of phase equilibrium in the liquid-vapour system with a single degree of freedom, pressure and temperature are independent parameters of equilibrium for sorption and the separation factor α can depend on the amount of sorbed gas. Therefore the concept of a differential separation factor, which characterizes the separation on a given portion of the isotherm of sorption, is introduced [3.3, 4]:

$$\alpha_{dif} = \left(\frac{P_1}{P_2}\right)_n, \tag{3.2}$$

where P_1 and P_2 are the equilibrium pressure of the pure components over a sorbent at its equal filling n.

Considering the filling of all the previous portions of the isotherm, also including the given one, and taking into account that hydrogen sorption is accompanied by its dissociation into atoms, one can derive the following expression for the separation factor

$$\ln \alpha_{A-B}^0 = \frac{1}{2n} \int_0^n \ln \alpha_{dif} dn = \frac{1}{2n} \int_0^n \ln \left(\frac{P_1}{P_2} \right)_n dn \tag{3.3}$$

where α_{A-B}^0 is the separation factor at equal ratio of hydrogen isotopes A and B (e. g., H and D) in the gas phase.

Using (3.3) and knowing the experimental isotherms of sorption, α can be found by graphical integration. When analytical expressions are available for the sorption isotherms, the calculation of α is simplified due to the possibility of calculation of the integral involved in (3.3). As shown in Chap. 2, the systems considered in this work are typified by irreversible processes of sorption–desorption on definite portions of the isotherms, which leads to the appearance of hysteresis on the isotherms and complicates the calculation of separation factors.

Let us consider the separation factors corresponding to three characteristic portions of the isotherms of sorption: α-phase, α–β-transition, and β-phase.

α-phase. In the range of low concentrations of hydrogen, where dissolving of atoms is not followed by any essential change in the crystal lattice, the dependence of the equilibrium pressure on the composition is described by Sieverts law:

$$\left(P_{H_2} \right)^{1/2} = K_{S,H_2} \cdot n, \tag{3.4}$$

where K_S is the Sieverts constant, n is the hydrogen concentration in the solid phase expressed as the atomic ratio H/Me.

From (3.3) we find that in the range of the α-phase the separation factor remains constant, independent of pressure

$$\alpha_{dif} = \alpha_{A-B}^0 = \frac{K_{S,A_2}}{K_{S,B_2}}. \tag{3.5}$$

α–β-transition. In this range the ratio of pressures of hydrogen isotopes over a wide range of fillings, as a rule, changes only slightly. If these changes are ignored, the differential separation factor is expressed as

$$\alpha_{dif,H-D} = \left(\frac{P_{A_2}}{P_{B_2}} \right)^{1/2}, \tag{3.6}$$

where P_{A_2} and P_{B_2} are equilibrium pressures corresponding to the horizontal portion of the sorption isotherms of hydrogen molecules A_2 and B_2 (e. g. isotherms of H_2 and D_2 for $\alpha_{dif,H-D}$).

In the case of moderate hydrogen filling of the α-phase, the separation factor α_{A-B}^0 can be estimated using (3.6). Table 3.1 presents the results of a calculation

Table 3.1. Determination of α^0_{H-D} from the isotherms of protium and deuterium sorption

IMC(Me)	Reference	T	α^0_{H-D}		
		K	calculation by (3.6)	graphical integration	experimental data
LaNi$_5$	[3.6]	218	1.21	1.25	1.22
TiMn$_{1.5}$	[3.7]	228	1.62	1.60	1.64
TiCrMn	[3.8]	228	1.80	1.80	1.70
TiMn$_{1.4}$Ni$_{0.1}$	[3.8]	250	–	1.50	–
V	[3.9]	313	1.73	–	–
Nb	[3.9]	333.6	1.15	–	–
Nb$_{0.2}$V$_{0.8}$	[3.9]	318.5	1.45	–	–
U	[3.10]	600	1.34	–	1.36

of α^0_{H-D} from the sorption isotherms of H_2 and D_2 for quite a number of systems. Values of P_{H_2} and P_{D_2} are taken at a filling $n = 1.5$ g-atom H(D)/g-atom Me.

When there are appreciable differences between separation factors in the α- and β-phases, the separation factor in the transition range can be found in terms of the number of hydrogen atoms in the two phases from the equation of additivity

$$\overline{\alpha}^0_{A-B} = \alpha^0_{A-B,\alpha}\varphi + \alpha^0_{A-B,\beta}(1 - \varphi), \tag{3.7}$$

where φ is the fraction of hydrogen atoms retained in the α-phase of the two-phase region in relation to the overall hydrogen content.

β**-phase.** This range is of prime interest for isotope separation and is typified by a variety of forms of $n = f(P)$. Thus graphical integration of the isotherms of sorption of isotope modifications of molecular hydrogen is the most general-purpose method for calculating α. The results of α^0_{H-D} calculation using (3.3) summarized in Table 3.1 agree well with the result of the direct experimental determination of the isotope effects.

3.2 Experimental Methods for Investigating Isotope Equilibrium

Adjusting a hydrogen isotope for equilibrium with a hydride-forming substance, as for other two-phase systems, is one of the most simple direct methods of determining the separation factor for isotope exchange between hydrogen gas and solid (hydride) phases. The method consists in establishing equilibrium between the phases, determining the equilibrium concentrations x and y within each phase, and the subsequent calculation of α using (3.1).

The experimental apparatus can be constructed both from glass, if isotope exchange is studied under pressures of no more than 1 atm [3.5] or from stainless steel [3.6]. In the latter case activation of metal or IMC can be performed directly in the plant for isotope equilibrium study.

Figure 3.2 presents the scheme of the metallic apparatus used for studying isotope exchange of hydrogen with hydride phases of IMC at pressures up to 3 MPa. The plant consists of the following main units: reactor (1) equipped with filter to avoid the entrainment of IMC particles; plunger circulatory pump (3) enabling one to control the velocity of gas flow in the range $0.2-1.0\,\mathrm{m^3/h}$; accumulators of hydrogen and its isotopes (9, 10); reference manometer (4); evacuation system and vacuum control.

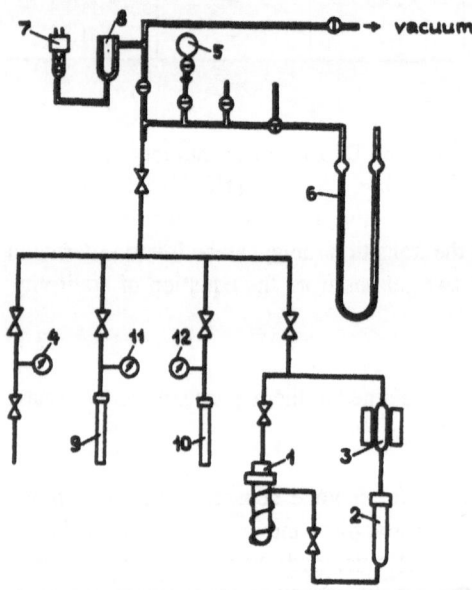

Fig. 3.2. Scheme of the plant for the investigation of hydrogen isotope exchange. (1) reactor, (2) buffer vessel, (3) circulatory pump, (4) reference manometer, (5) container for sampling, (6) mercury manometer, (7) thermocouple lamp PMT-2, (8) trap with liquid nitrogen, (9, 10) hydrogen accumulators, (11, 12) manometers on hydrogen accumulators

Once the IMC samples has been evacuated, the reactor is filled with hydrogen of definite isotope composition. When sorption is completed, the reactor is disconnected from the rest of the circulatory loop, which is filled with hydrogen of another isotope composition at a pressure equal to that in the reactor. Once the reactor has been connected the circulatory pump is switched on and the change in concentration of the heavy isotope (deuterium or tritium) is monitored by periodic sampling. Another way of monitoring the concentration of the heavy isotope is to equip the setup with a special detector that makes it possible to carry out continuous measuring of the isotopic composition of the gas. The following devices can be used for this purpose: a mass-spectrometer to measure the deuterium concentration and an ionization chamber for measuring the tritium concentration.

When the equilibrium isotope composition in the phases has been established, the working volume is disconnected from the remainder of the circulatory loop and thermal desorption of hydrogen from metal or IMC is carried out. As a rule, desorption is performed into the circulatory loop initially freed from gas. Since a kinetic isotope effect may occur during desorption, the gas released is carefully mixed in the loop by the circulatory pump and after this step the isotope compostion of the gas released is determined.

When calculating the separation factor it is necessary to introduce a correction for the mixing of the gas released with hydrogen of equilibrium isotope composition, which remains in the free space of the reactor (1) before desorption.

To check the degree to which equilibrium has been attained additional experiments are carried out, in which the sequence of bleeding the mixtures of different isotope composition in the reactor and the rest of the circulatory loop is reversed.

To provide constant temperature the gas is first heated by passing through the snake-shaped heat exchanger, which is thermostated together with the reactor. In the case of periodic control of isotope composition the gas sampling is performed with no marked disturbance of the phase and isotope equilibrium. For a gas sample of volume up to $1\,cm^3$ s.t.p. it appears to be sufficient for the metal or IMC to have a mass 5–10 g.

The most reliable results are obtained when it is possible to check the mass balance both as a whole for hydrogen and for the particular isotope. It is evident that the reliability of such a check is higher when the ratio of hydrogen in the two phases is close to 1.

Studies of isotope equilibrium in the range of the α-phase fulfilling the mentioned demands are impeted by the low content of hydrogen in the solid phase. This frequently results in a decrease of the accuracy in determining α.

The difficulty of measuring α at the rigorously predetermined isotope composition of a phase (for example, when determining α_{H-D}^0) is another drawback of the method described, since during the process of single adjusting for equilibrium, the compositions of the phases change. It is also to be noted that in simultaneous studies of the equilibrium of all three hydrogen isotopes with hydride phases it is necessary to apply different methods of analysis, which causes a different accuracy in the determination of α for the binary mixtures H–T, H–D, and D–T. Only the method of laser desorption is free from drawbacks [3.11]. Let us illustrate its application by the example of the systems Pd–$H_2(D_2)$ and TiMn$_{1.5}$–$H_2(D_2)$. To determine the separation factors requires a minor amount of the metal or IMC (of the order of 1 mg), which is adjusted for equilibrium at the corresponding temperature with a mixture of D_2 and H_2 in the ratio 1 : 1 (at a total pressure of 1.010 mbar, the required amount of the gas phase is 1–3 cm^3 s.t.p.). Since the ratio of hydrogen in hydride and gas phases $\lambda \ll 1$, the isotope composition of the gas phase during the process of equilibrium establishment is practically constant. This enables one to simplify (3.1) by transformation to the form: $\alpha_{H-D}^0 = x/(1 - x)$.

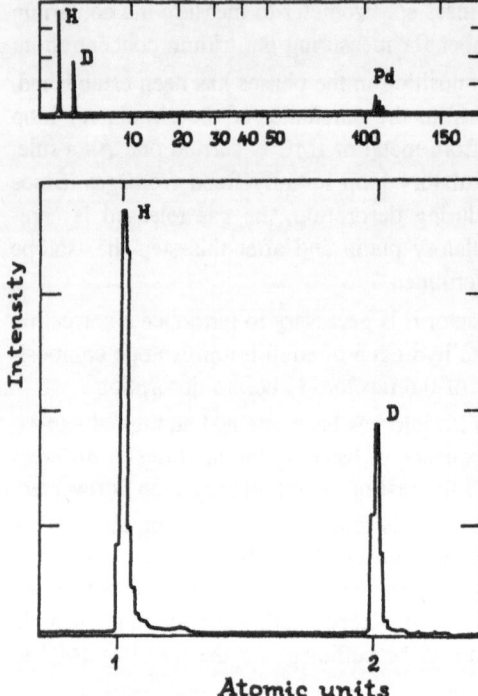

Fig. 3.3. Mass spectra of palladium hydride by laser desorption: top – total spectrum; bottom – magnified fragment of the spectrum

So, to find α_{H-D}^0 it is necessary to know the ratio of deuterium and protium concentrations in the hydride phase; this can be found by means of the laser desorption.

To carry out the experiments one needs a modified laser mass-spectrometer LAMMA-1000 making it possible to cool the sample placed in the device down to 80–90 K, i. e., down to such a temperature, at which appreciable decomposition of hydrides in vacuum is not observed.

Preliminary conditions of preparation of the samples enable to synthesize β-hydride of Pd and α-phase of TiMn$_{1.5}$ hydride (at higher pressures the data for β-phase of TiMn$_{1.5}$ hydride also can be obtained).

The spectrum is taken in the range 0–150 Dalton. The depth of the crater formed by a single laser shot is equal to $\sim 0.1\,\mu$m. To avoid possible error in the determination of α_{H-D}^0 due to the influence of chemisorbed water and possible evacuation of protium and deuterium from the surface of hydrides, the spectrum is taken after each tenth shot at ten different points on the sample.

Figures 3.3, 4 summarize the spectra of the positive ions for the two samples explored and show magnified protium and deuterium peaks for Pd hydride, and magnified Ti- and Mn-peaks for TiMn$_{1.5}$. The ratio corresponding to the separation factor α_{H-D}^0 is found from the ratio of the areas under H$^+$ and D$^+$ peaks. These values are equal to 0.42 for Pd and 1.14 for TiMn$_{1.5}$ at 298 K, which agrees well with the data obtained in [3.12, 13], namely 0.41 and 1.18, respectively.

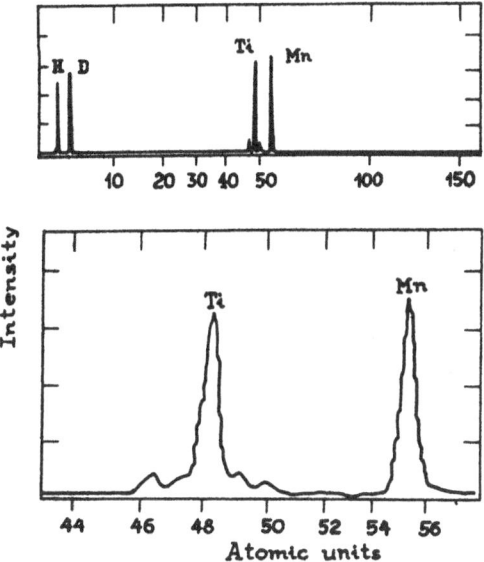

Fig. 3.4. Mass spectra of $TiMn_{1.5}$ hydride by laser desorption: top – total spectrum; bottom – magnified fragment of the spectrum

It is note worthy that such a simple technique of α_{H-D}^0 calculation is appropriate only in the case of the absence of ions H_2^+, HD^+, and D_2^+ as well as ions OH^+, OD^+. A check on the formation of MeH^+ ions is also important.

As is evident from Figs. 3.3, 4, in the range 2–40 Dalton the peaks corresponding to the masses of the above-mentioned ions are absent. Figure 3.4 also shows that peaks for the ions TiH^+, TiD^+, MnH^+, MnD^+ are absent and only mass peaks 46, 47, 48, 49, 50 are observed for the isotopes of Ti^+ and 55 for Mn^+.

The method of laser desorption enables one to simultaneously check the IMC composition. This makes it possible to determine the influence of IMC composition on the value of the separation factor. The method also enables one to find the factors α_{D-T}^0 and α_{H-T}^0 using minor amounts of tritium (~ 2.5 Ci) and α_{H-D}^0 at high pressure.

The drawback of this method is the necessity to use the extremely expensive device LAMMA-1000.

Apart from the method of single adjusting for equilibrium, the chromatographic method is also used to determine the separation factor in systems with hydride phases of metals and IMC. Front and distributive chromatography are the most frequently used variants of this method. In the first case, the outlet fronts of the isotopes provide information for correlating their behaviour in the column. The latter information and the known isotopic concentration of the source mixture makes it possible to calculate the separator factor.

To find α by the method of distributive chromatography one measures the time of output of each peak whose number corresponds to one of the components extracted from the starting mixture. The value of α is computed as the ratio of output times of the corresponding peaks. When systems with a hydride phase of metals or IMC are used, it should be taken into account that, as a rule, on the surface of the solid phase a homomolecular isotope exchange reaction of hydrogen occurs (for example, $H_2 + D_2 \leftrightarrow 2HD$). Thus the distribution of isotopes between the isotope modifications of hydrogen molecules differs from the distribution in the source gas. In this case isotope modifications of hydrogen molecules not found in the source mixture can arise (e. g., HD). The possibility of determining the separation factors for all binary mixtures of hydrogen isotopes in a single experiment is the merit of this method.

As is evident from an analysis of published data, values of α determined by the chromatographic method presented in different works differ from each other and from the values obtained by the method of single adjusting for equilibrium.

The causes of such discrepancies can be related to the common drawback of the chromatographic method for α determination, namely the difficulties in determining the degree of filling of the solid phase (the composition of the solid phase to which the derived value of α corresponds). Another basic drawback is that the thermodynamic characteristic – separation factor – is found under unsteady conditions. That is why, in the chromatographic process, additional separation of components caused by kinetic isotope effects can occur (e. g., separation of components of a mixture during desorption). Finally, when carrying out separation in highly effective short columns, errors may arise due to the neglect of the influence of the injected sample volume on the output time of the peak.

3.3 Calculation of Separation Factors by the Harmonic Oscillator Model. Positive and Negative Isotope Effects

The quantum-statistical method is widely used for the calculation of thermodynamic isotope effects, particularly for chemical isotope exchange reactions occurring in the gas phase. In the last case, for known vibrational frequencies of isotopic molecular modifications of the exchanging molecules the accuracy of quantum-statistical calculations of isotope equilibrium is high and usually exceeds the accuracy of experimentally obtained separation factors. Since isotopic frequency shifts $\Delta\omega$ of any isotope modification of molecular hydrogen are determined with a high degree of accuracy, the basic difficulties are concerned with the calculation of the partition function of the solid phase.

Considering isotope exchange of gaseous hydrogen with the hydride phase as a routine reaction of chemical isotope exchange:

$$2A(Me) + B_2 \leftrightarrow 2B(Me) + A_2, \tag{3.9}$$

it is possible to write the following expression for the equilibrium constant of the reaction:

$$K_{A-B} = \frac{Z_{A_2}}{Z_{B_2}} \cdot \frac{Z_{B(Me)}^2}{Z_{A(Me)}^2} \tag{3.10}$$

where Z_{A_2} and Z_{B_2} are the partition functions of hydrogen molecules with light and heavy isotopes, respectively (e. g., for isotope exchange H–T it means Z_{H_2} and Z_{T_2}), $Z_{A(Me)}$ and $Z_{B(Me)}$ are the partition functions of the hydride phases of metal or IMC including light and heavy hydrogen atoms (for the example considered $Z_{H(Me)}$ and $Z_{T(Me)}$).

Considering the known relation of K with α [3.14, 15]

$$\alpha^0_{A-B} = (K_{A-B})^{1/2} \tag{3.11}$$

one can obtain

$$\alpha^0_{A-B} = \left(\frac{Z_{A_2}}{Z_{B_2}}\right)^{1/2} \left(\frac{Z_{B(Me)}}{Z_{A(Me)}}\right). \tag{3.12}$$

A peculiarity of hydrogen isotope modifications is their non-equiprobable distribution in the reactions of homomolecular isotope exchange (HMIE)

$$H_2 + T_2 = 2HT, \quad K_{HT}, \tag{3.13}$$
$$H_2 + D_2 = 2HD, \quad K_{HD}, \tag{3.14}$$
$$D_2 + T_2 = 2DT, \quad K_{DT}, \tag{3.15}$$

or in general form

$$A_2 + B_2 = 2AB, \quad K_{AB}. \tag{3.16}$$

This implies $Z_{A_2}/Z_{AB} < Z_{AB}/Z_{B_2}$, whereas, for diatomic molecules not containing hydrogen, the equalities

$$\frac{Z_{A_2}}{Z_{AB}} = \frac{Z_{AB}}{Z_{B_2}} = \left(\frac{Z_{A_2}}{Z_{B_2}}\right)^{1/2} \tag{3.17}$$

are rigorously obeyed.

Since violation of the equalities (3.17) in systems with molecular hydrogen leads to a concentration dependence of the separation factor as considered in Sect. 3.4 it should be useful to distinguish the processes of isotope exchange in the range of low

$$A(Me) + AB = B(Me) + A_2 \tag{3.18}$$

and high

$$A(Me) + B_2 = B(Me) + AB \tag{3.19}$$

content of the heavy isotope B.

Calculation of the partition functions for isotope modifications of molecular hydrogen can be performed conveniently using the expression

$$\ln\left(\frac{\sigma_{AB} Z_{AB}}{\sigma_{H_2} Z_{H_2}}\right) = \sum_{n=0}^{4} a_n \left(\frac{300}{T}\right)^n \tag{3.20}$$

obtained by *Bron* with co-workers [3.16] from spectral data. The $\sum_{n=0}^{4}$ coefficients of this polynomial for all the hydrogen isotope modifications are presented in Table 3.2.

Table 3.2. Coefficients of (3.20)

A	B	a_0	a_1	a_2	a_3	a_4
H	D	0.89156	1.36456	−0.00075	−0.00625	0.00638
H	T	1.44049	1.85581	0.02076	−0.02085	0.01015
D	D	1.74669	2.88780	0.16472	−0.12030	0.03493
D	T	2.26697	3.49816	0.18639	−0.12791	0.03531
T	T	2.78976	4.07518	0.37271	−0.24982	0.06440
K_{HT}		1.47751	−0.36356	−0.33119	0.20811	−0.04410
K_{HD}		1.42272	−0.15868	−0.16622	0.11405	−0.02177
K_{DT}		1.38378	0.00334	−0.16465	0.11429	−0.02871

A diversity of works [3.6, 8, 12, 13, 15, 17–22] presents moderate agreement of the experimental values of α obtained in a wide temperature range (195–353 K) with the values computed by the quantum-statistical method using a three-dimensional harmonic oscillator model for the hydride phase. According to the model the ratio of the partition functions of heavy and light hydrogen isotopes in the crystal lattice of the metal or IMC is related to the frequencies of three-fold degenerate atomic modes ω_A and ω_B by the equation

$$\frac{Z_{B(Me)}}{Z_{A(Me)}} = \left(\frac{1 - \exp(-\hbar\omega_A/kT)}{1 - \exp(-\hbar\omega_B/kT)} \exp\left[\frac{\hbar(\omega_A - \omega_B)}{2kT} \right] \right)^3 , \tag{3.21}$$

or

$$\frac{Z_{B(Me)}}{Z_{A(Me)}} = \left(\frac{\sinh(u_A/2)}{\sinh(u_B/2)} \right)^3 , \tag{3.22}$$

where $u_A = \hbar\omega_A/kT$ and $u_B = \hbar\omega_B/kT$.

Since for a harmonic oscillator $\omega_H = \omega_D 2^{1/2} = \omega_T 3^{1/2}$, performing the calculations of the separation factors for the mixtures H–T, H–D, and D–T and their temperature dependencies requires the single value ω, which can be found from the single experimental value of α for any mixture.

For the reactions (3.18) and (3.19), in which symmetric hydrogen molecules transform to asymmetric ones (and conversely), the separation factor is related to the constant of equilibrium by

$$\alpha = \frac{K}{K^\infty} , \tag{3.23}$$

where K^∞ is determined by the ratio of symmetry numbers of hydrogen molecules involved in the reaction.

For reaction (3.18) the equality $K^\infty = \sigma_{AB}/\sigma_{A_2} = 1/2$ is valid and for the reaction (3.19) we have $K^\infty = \sigma_{B_2}/\sigma_{AB} = 2$.

So the separation factor is found to be equal to

$$\alpha = \left(\frac{\sin \hbar(u_A/2)}{\sin \hbar(u_A/2(m_B/m_A)^{1/2})}\right)^3 / \exp\left[\sum_{n=0}^{4} a_n \left(\frac{300}{T}\right)^n\right]. \tag{3.24}$$

When performing quantum-statistical calculations of thermodynamic isotope effects the concept of β-factor introduced by *Warshawskii* and *Waysberg* [3.23] appears to be useful. It manifests itself in the equilibrium constant of the isotope exchange reaction of a molecule with an atom. For the hydride phase, the β-factor is equal to the equilibrium constant of the reaction

$$A(Me) + B^\bullet \leftrightarrow B(Me) + A^\bullet \tag{3.25}$$

defined by the relation

$$K = \frac{Z_{B(Me)} Z_{A^\bullet}}{Z_{A(Me)} Z_{B^\bullet}}. \tag{3.26}$$

It is expressed by the equality

$$\beta_{Me(A-B)} = \left(\frac{m_A}{m_B}\right)^{3/2} \left[\frac{\sinh(u_A/2)}{\sinh(u_A/2(m_B/m_A)^{1/2})}\right]^3 \tag{3.27}$$

where $m_A/m_B = Z_{A^\bullet}/Z_{B^\bullet}$ is equal to the ratio of masses of atoms of light and heavy hydrogen isotopes.

In the case of molecular hydrogen, for any pair of isotopes three values of the β-factor can be considered, which differ markedly from each other, especially at low temperatures:

– In the range of low content of the heavy isotope B

$$\beta_{A_2-AB} = \frac{u_{AB}}{u_{A_2}} \frac{\sinh(u_{A_2}/2)}{\sinh(u_{AB}/2)}; \tag{3.28}$$

– In the range of high content of the heavy isotope

$$\beta_{AB-B_2} = \frac{u_{B_2}}{u_{AB}} \frac{\sinh(u_{AB}/2)}{\sinh(u_{B_2}/2)}; \tag{3.29}$$

– At equal content of light and heavy isotopes

$$\beta_{A_2-B_2}^0 = \left(\frac{u_{B_2}}{u_{A_2}} \frac{\sinh(u_{A_2}/2)}{\sinh(u_{B_2}/2)}\right)^{1/2}. \tag{3.30}$$

In these experession the following symbols are used:

$$u_{A_2} = \frac{\hbar\omega_{A_2}}{kT}, \quad u_{AB} = \frac{\hbar\omega_{AB}}{kT}, \quad \text{and} \quad u_{B_2} = \frac{\hbar\omega_{B_2}}{kT}.$$

As will be shown below, the separation factors and hence β-factors of hydrogen in the range of low and high isotope contents are related to the constants of equilibrium of the HMIE reactions (3.13–16). The equilibrium constants of all the three HMIE reactions can be found using (3.20) for the different isotope modifications of hydrogen molecules. The temperature dependence of K_{AB} is expressed by the equation

$$\ln K_{AB} = \sum_{n=0}^{4} a_n \left(\frac{300}{T} \right)^n , \qquad (3.31)$$

whose coefficients are presented in Table 3.2. The temperature dependence of the β-factors of hydrogen for the mixtures H–T and D–T for both low and high contents of the heavy isotope are shown in Fig. 3.5.

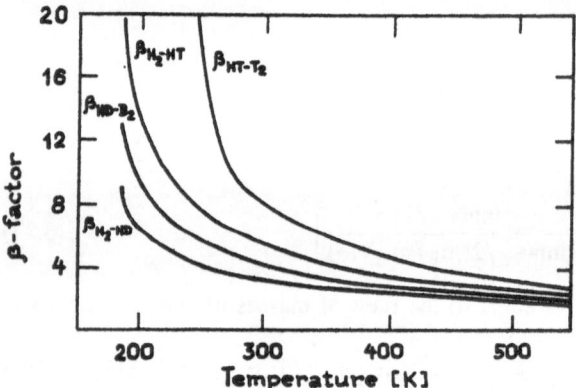

Fig. 3.5. Temperature dependencies of β-factors of molecular hydrogen at exchange H–T ($\beta_{H_2-HT}, \beta_{HT-T_2}$) and H–D ($\beta_{H_2-HD}, \beta_{HD-D_2}$)

The value of the β-factor is a quantitative measure of the inequality of two isotopes of the same element in the compound under consideration. If isotope atoms are found to be identical the β-factor is equal to 1. The increase of the β-factor points to an increasing change in the quantum-mechanical characteristics upon isotope substitution in molecules.

Figure 3.6 shows the dependence of the β-factor for isotope substitution H–T on $\hbar\omega$ computed for the hydride phase using (3.27) at temperatures of 173 and 273 K. Independently of temperature, at $\hbar\omega = 0$, $\beta_{Me(A-B)} = 1$; it increases relatively weakly with increasing frequency of local modes up to 60 meV at $T = 173\,K$ and up to 80 meV at $T = 273\,K$. Then one observes an abrupt (practically exponential) increase of the β-factor as the harmonic oscillator frequency increases. The ranges of concentration change of the hydrogen β-factors at the two temperatures considered, whose bounds are defined by (3.28, 29), are shown as shaded regions. At low values of the local mode frequency, the heavy isotope is concentrated in the gas phase. At certain values of the local mode frequency depending on temperature

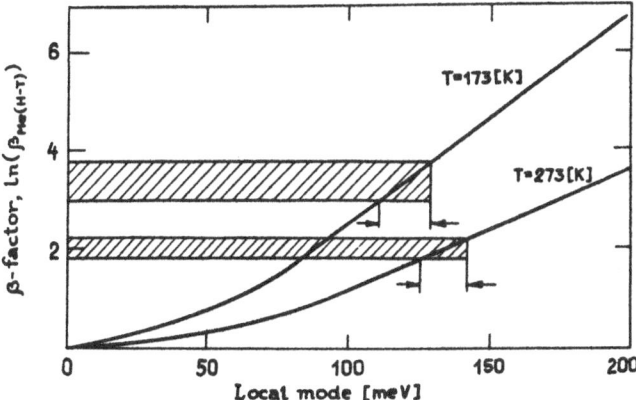

Fig. 3.6. Dependencies of $\ln \beta_{Me(H-T)}$ on local mode energy at temperatures of 173 and 273 K

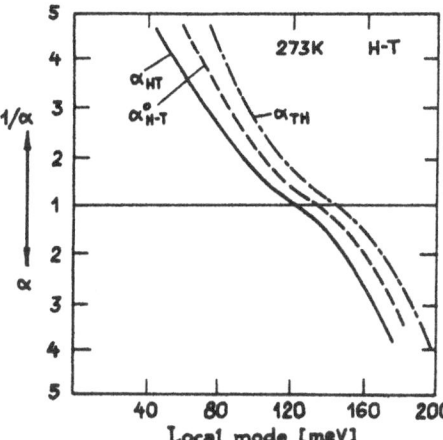

Fig. 3.7. Dependencies of α_{HT}, α_{H-T}^0, and α_{TH} on vibrational energy of harmonic oscillator at 273 K

and concentration range of isotope separation, an inversion of the isotope effect is observed and thereafter the gas phase will be enriched with the light isotope, i. e. a positive isotope effect will be observed. At a temperature of 173 K the local mode of inversion ranges from 110 meV (at low tritium concentration) to 130 meV (at high tritium concentration). The local mode of inversion increases with increasing temperature and at 273 K this zone of inversion is localized within the range 125–145 meV. The difference of ordinates in Fig. 3.6, i. e. the distance between the horizontal straight line characterizing the value of the hydrogen β-factor and the curve $\ln \beta_{Me(H-T)} = f(\hbar\omega)$ is equal to $\ln \alpha$. The dependence of α on the local mode energies for $T = 273$ K is presented in Fig. 3.7. The lower curve characterizes the isotope effect in the range of low tritium content; the middle curve is obtained for a ratio of isotopes in the gas phase of H : T = 1 : 1, and the upper curve refers to the case of high tritium content. Maximal values of α for the negative isotope effect correspond to $\hbar\omega = 0$ and are equal to the value of the hydrogen β-factor according to (3.28–30).

Figure 3.8 shows the dependences of hydride phase β-factors for all three pairs of isotopes on the value of the reduced temperature $u = \hbar\omega/kT$, from which the value of the β-factor can be found for the range of temperatures and frequencies of the harmonic oscillator local mode of practial interest. Each of these curves can be interpreted not only as the dependence of the β-factor on ω at $T = $ const but also as its dependence on $1/T$ at $\omega = $ const. Thus, over a wide temperature range, $\ln\beta$ depends linearly on $1/T$. Only in the range of high temperatures are deviations from the linear dependence observed (for the H–T mixture the deviations appear at a lower temperature than for the D–T and H–D mixtures). A decrease of ω also leads to a decrease of the limiting temperature still corresponding to the linear dependence in Fig. 3.8.

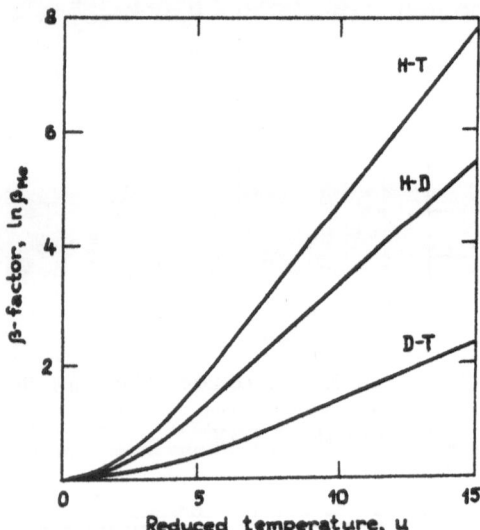

Fig. 3.8. Dependencies of $\ln\beta_{Me(H-T)}$, $\ln\beta_{Me(H-D)}$, and $\ln\beta_{Me(D-T)}$ on the reduced temperature u

Let us analyze possible values of the separation factors and their temperature dependences at equilibrium of hydrogen with hydride phases. Figure 3.9, 10 show the temperature dependences of α_{HD} and α_{HT} calculated from (3.24) in the range of low heavy isotope content, i. e. corresponding to the reaction (3.18) [3.21]. The largest separation factors for the negative isotope effect are limited by the upper curves in the figures, which correspond to hydride phase β-factors equal to 1. The negative isotope effect decreases up to the inversion with increasing local mode energy.

At values of local mode energies around 100 meV an extreme character of the temperature dependence of α is revealed. In case of HMIE reactions such behaviour is not known. As is evident from Figs. 3.9, 10, even in systems with considerable positive isotope effect, its inversion can be observed at high tem-

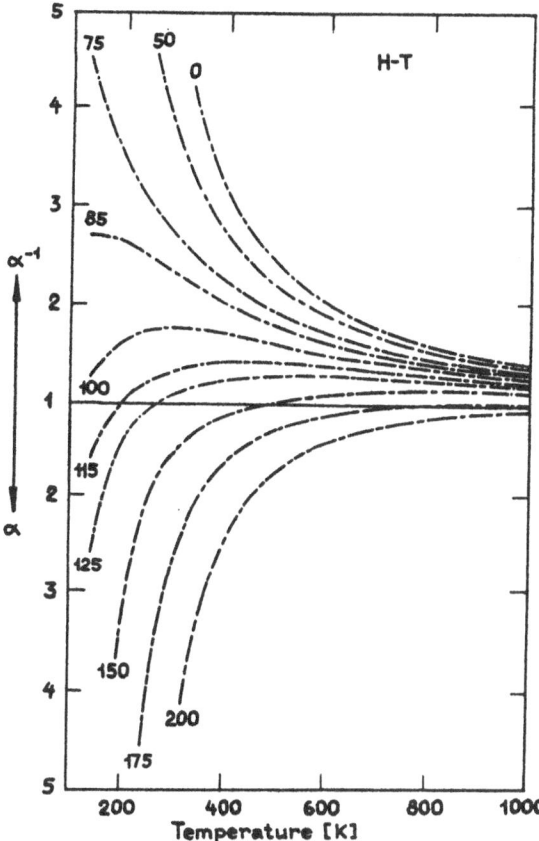

Fig. 3.9. Temperature dependencies of the separation factor α_{HT} at different local mode energies of the hydrogen atom in the hydride phase

peratures. Figure 3.11 presents the dependence of the isotope effect inversion temperature on the value of local mode energies for the mixtures H–D and H–T, whence it follows that the inversion temperature depends weakly on the type of isotope mixture (it is approximately equal for the mixtures H–D and H–T).

The above-performed analysis is based on the harmonic oscillator model with three degenerate frequencies for computing the partition function of the hydride phase. As will be shown below, in a number of real systems the energy of an atom in the crystal lattice of the hydride phase can be characterized by two or three local-mode frequencies. In this case the use of the harmonic oscillator model results in the following expressions for the ratio of the partition functions and β-factors of the hydride phase:

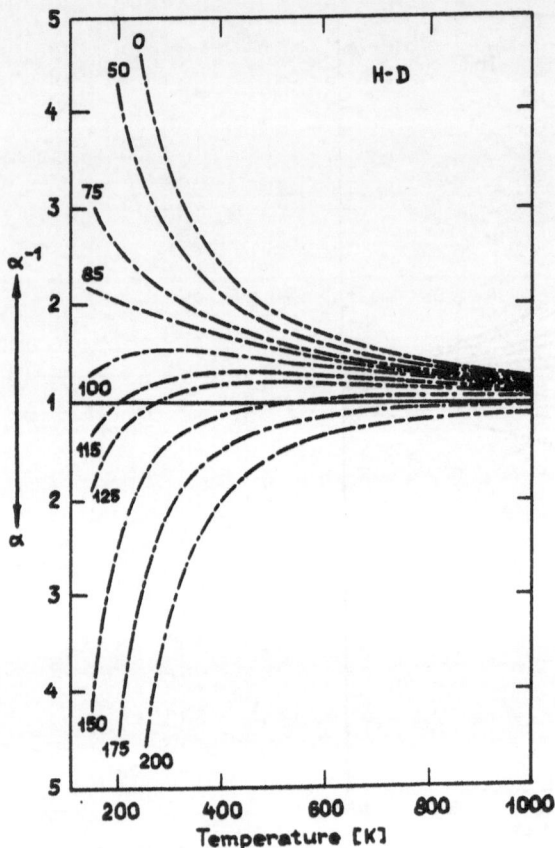

Fig. 3.10. Temperature dependencies of the separation factor α_{HD} at different local mode energies of the hydrogen atom in the hydride phase

– For the ratio of the partition functions

$$\frac{Z_{B(Me)}}{Z_{A(Me)}} = \prod_{i=1}^{3} \left[\frac{1 - \exp(-u_{A,i})}{1 - \exp(-u_{B,i})} \exp\left(\frac{u_{A,i} - u_{B,i}}{2} \right) \right] \tag{3.32}$$

or

$$\frac{Z_{B(Me)}}{Z_{A(Me)}} = \prod_{i=1}^{3} \frac{\sinh(u_{A,i}/2)}{\sinh(u_{B,i}/2)} ; \tag{3.33}$$

– For the β-factor of the hydride phase

$$\beta_{Me(A-B)} = \left(\frac{m_A}{m_B} \right)^{3/2} \prod_{i=1}^{3} \frac{\sinh(u_{A,i}/2)}{\sinh(u_{A,i}/2(m_B/m_A)^{1/2})} . \tag{3.34}$$

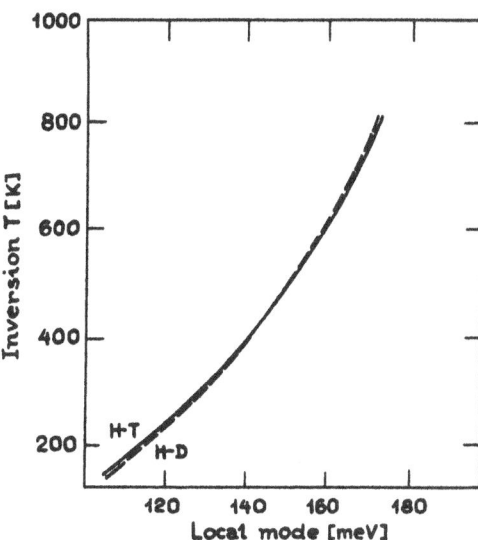

Fig. 3.11. Dependence of the separation effect inversion temperature in H–T and D–T mixtures on the vibrational energy of the local mode

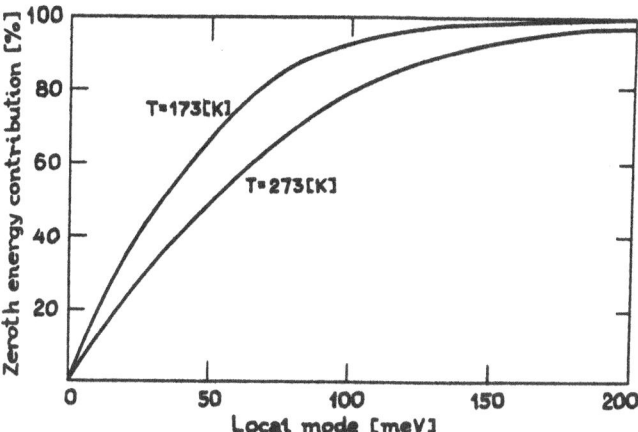

Fig. 3.12. Dependence of the zero-point energy contribution (in %) to $Z_{T(Me)}/Z_{H(Me)}$ or $\beta_{Me(H-T)}$ on vibrational energy at temperatures of 173 and 273 K

As follows from Fig. 3.12, the contribution of excited levels to the partition function or β-factor of hydrogen atoms in the hydride phase in systems with a positive isotope effect at $T \leq 273$ K is slight and, hence, the foregoing equations can be simplified. For example, (3.32) can be written as

$$\frac{Z_{B(Me)}}{Z_{A(Me)}} = \exp\left(\sum_{i=1}^{3} \frac{u_{A,i} - u_{B,i}}{2} \right), \tag{3.35}$$

So, the calculation of (3.32, 33) reduces to finding the partition function or β-factor for a harmonic oscillator with three frequencies, whose value is equal to

$$u_A = \frac{1}{3 \sum_{i=1}^{3} u_{A,i}}. \qquad (3.36)$$

The model with two parameters, which was also used when calculating the isotope effects in hydrogen–metal systems [3.24, 25], should also be mentioned. However, a great amount of information is require for calculations using this model.

3.4 Dependence of the Separation Factor on Isotope Concentration

As noted above, the deviation of the equilibrium distribution of isotopes from the equiprobable one in the HMIE reactions causes the dependence of the separation factor on concentration [3.14]. In all systems typified by chemical exchange of hydrogen with molecules of other substances and containing two hydrogen atoms and more, the violation of the equiprobable distribution in the HMIE reactions leads to a partial compensation of the effect. This is because K_{ABX} (or K_{ABY}) is not equal to the limiting value K_{ABX}^{∞} (K_{ABY}^{∞}) corresponding to the equiprobable distribution ($K_{ABX} < K_{ABX}^{\infty}$ or $K_{ABY} < K_{ABY}^{\infty}$) [3.15, 26]. The concentration dependence appears to be strongest in isotope exchange processes between hydrogen gas and materials that contain molecules with only one hydrogen atom, and with hydride phases of metals and IMC, since these phases contain hydrogen in the atomic state.

If the HMIE reaction does not occur in a system, only limiting separation factors can be realized in practice: α_{A_2-AB} and α_{AB-B_2} in the range of low and high content of isotope B, respectively. This is observed for instance at equilibrium of hydrogen with zeolites, activated carbons, silica gels, and other sorbents of molecular hydrogen [3.8, 15]. In all cases, equilibrium of hydrogen with hydride phases of metals and IMC is accompanied by dissociation of hydrogen molecules and the HMIE reactions also occur on the solid surface with a rate which, as a rule, is several orders greater than the rate of interphase isotope exchange (IIE) between gaseous hydrogen and the solid phase. This leads to a continuous concentration dependence of α.

The concentration dependence of the separation factor means that there is still an isotope effect observed in the areas of high and low isotope concentrations, even if the sorption isotherms of both isotopes coincide (α_{A-B}^{0}).

Let us consider the concentration dependence of α for the positive isotope effect. The equilibrium constants of the reactions (3.18, 19) are respectively

$$K_1 = \frac{[B(Me)]\,[A_2]}{[A(Me)]\,[AB]}, \qquad (3.37)$$

$$K_2 = \frac{[B(Me)]\,[AB]}{[A(Me)]\,[B_2]}. \qquad (3.38)$$

Considering that for reaction (3.18) the separation factor $\alpha_{AB} = K_1/K_1^{\infty} = 2K_1$, for reaction (3.19) the separation factor $\alpha_{AB} = K_2/K_2^{\infty} = K_2/2$, and the HMIE reaction (3.16) can be derived by subtraction of reaction (3.18) from reaction (3.19), one can connect the limiting separation factors in the range of low and high heavy isotope content

$$\alpha_{BA} = \frac{\alpha_{AB} K_{AB}}{4}. \tag{3.39}$$

For the isotope mixture H–D, the relation between limiting separation factors α_{HD} and α_{DH} is considered in [3.27] for isotope exchange of hydrogen with the hydride phase of the alloy Pd–Pt. Subsequently, with the use of relation (3.39), the temperature dependence of α_{DH} is computed in the H_2–Pd system using the experimental values of α_{HD} and K_{HD} of the HMIE reaction [3.28].

Let us express the separation factor at any isotope composition in terms of α_{AB} starting from the definition of α according to relation (3.1)

$$\alpha_{A-B} = \frac{[B(Me)]}{[A(Me)]} \frac{2[A_2] + [AB]}{2[B_2] + [AB]} \tag{3.40}$$

or

$$\alpha_{A-B} = \frac{\alpha_{AB}}{2} \frac{2 + [AB]/[A_2]}{2[B_2]/[AB] + 1}. \tag{3.41}$$

Considering that

$$\frac{[B_2]}{[AB]} = \frac{[AB]}{K_{AB}[A_2]}, \tag{3.42}$$

one can derive the following expression for the concentration dependence of α:

$$\alpha_{A-B} = \alpha_{AB} \frac{1 + 2[A_2]/[AB]}{4/K_{AB} + 2[A_2]/[AB]}. \tag{3.43}$$

This expression is the simplest special case of the concentration dependence of the separation factor, including the HMIE reactions in both phases, which are in equilibrium [3.15, 26]. In the considered systems with hydride phases, the mathematical description of the concentration dependence of α is greatly symplified, because the hydrogen dissolved is dissociated into atoms.

At the given hydrogen isotope composition (e. g., at atomic fraction y of isotope B) the ratio of isotope modifications of hydrogen $[A_2]/[AB]$ can be found from the quadratic equation

$$y\left(\frac{[A_2]}{[AB]}\right)^2 - \left(\frac{1}{2-y}\right)\frac{[A_2]}{[AB]} - \frac{1-y}{K_{AB}} = 0, \tag{3.44}$$

which is obtained from the expression

$$y = \frac{[B_2] + 0.5[AB]}{[A_2] + [AB] + [B_2]} \tag{3.45}$$

by inserting the ratio $[B_2]/[AB] = [AB]/(K_{AB}[A_2])$ into it.

In the limiting case of $[AB]/[A_2] \to 0$, we have $\alpha_{A-B} \to \alpha_{AB}$ and for $[A_2]/[AB] \to 0$, $\alpha_{A-B} \to \alpha_{BA}$. For equal content of isotopes A and B in the gas phase ($y = 0.5$), in terms of the ratio $[A_2]/[AB] = (K_{AB})^{-1/2}$, we obtain

$$\alpha_{A-B}^0 = \frac{\alpha_{AB}(K_{AB})^{1/2}}{2} = (K_{A-B})^{1/2}, \tag{3.46}$$

which agrees with (3.11).

At a given isotope composition of the hydride phase (concentration x) the concentration dependence of the separation factor can be expressed as follows:

$$\alpha_{A-B} = \alpha_{AB} \frac{1 + (1-x)/(x\alpha_{AB})}{4/K_{AB} + (1-x)/(x\alpha_{AB})}. \tag{3.47}$$

The convenience of this equation is concerned with the fact that one need not solve the quadratic equation (3.44). Since $\alpha_{AB}x/(1-x) = [AB]/2[A_2]$, equation (3.47) transforms to (3.43). According to these relations, the separation factor decreases with increasing heavy isotope concentration. In systems with a negative isotope effect ($\alpha < 1$) this means an increase in the separation effect with increasing heavy isotope concentration. Such behaviour is an important advantage of such systems when they are used for the final stage of concentration processes of tritium.

Experimental studies of the concentration dependence of α have been performed more fully for the H_2–Pd system, which is typified by the negative isotope effect. Figure 3.13 shows the experimental values of α_{H-D}^{-1} obtained by the method of single adjusting for equilibrium at equilibrium deuterium contents in the gas phase from 2 to 98 at% [3.12] and those obtained experimentally in [3.29, 30]. By linear extrapolation of the dependence in the range of low deuterium content the limiting value $\alpha_{HD}^{-1} = 2.16$ is found. This coincides with the value obtained by carrying out the experiments with hydrogen of the native isotope composition [3.31]. The concentration dependence of α_{H-D}^{-1} computed using this value with (3.43) is presented in Fig. 3.13 as a solid line. The calculated dependences obtained in a similar way for the mixtures H–T and D–T are also presented in Fig. 3.13 as solid lines. The initial data for calculation are the following: $\alpha_{HT}^{-1} = 2.93$ (the average of 3.03 [3.32] and 2.84 [3.33]) and $\alpha_{DT}^{-1} = 1.47$ [3.32]. For the H–T mixture, the experimental values of α_{H-T} obtained at equilibrium concentrations of tritium in the gas phase of 89 and 73 at% and at a temperature of 293 K, are also shown [3.34].

Furthermore Fig. 3.13 presents the experimental data [3.10, 35] and the calculated curve [3.22] for the separation factor α_{H-D} at equilibrium of hydrogen with the hydride phase of uranium at temperatures of 500–600 K.

When comparing the experimental values of α_{H-D} obtained by the method of single adjusting for equilibrium with the ratio of equilibrium pressures of the gas over β-phase hydride and deuteride of palladium [3.29], we must take into account the effect of isotope composition of the hydride phase on the value of the thermodynamic isotope effect and derive the following equation relating α to $\gamma = (P_{D_2}/P_{H_2})^{1/2}$:

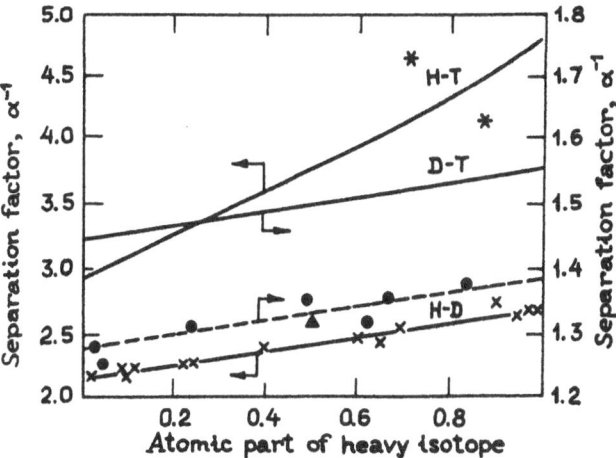

Fig. 3.13. Dependence of the separation factor in the systems H_2–Pd and H_2–U (dotted curve) on composition of the gas phase: • data from [3.28], ▲ data from [3.30], * data from [3.34]

$$\frac{1}{\alpha\gamma} = 1 + \frac{x(y+1)}{x\left[\gamma(K_{HD}/4)^{1/2} - 1\right] + 1}\left[1 - (K_{HD}/4)^{1/2}\right] \tag{3.48}$$

where P_{D_2} and P_{H_2} are the equilibrium pressures over the hydride and deuteride of palladium in the range of the β-phase at equal ratio n.

After analysis of the effect of the second term on the right-head side of (3.48), which in the experimental conditions ($x \approx 0.4$) appears to be relatively minor, the authors of [3.29] assumed that $\alpha^{-1} = \gamma$ and obtained the temperature dependence of α, which, in the authors' opinion, agrees satisfactorily with the experimental data. The authors associated the appreciable discrepancies (e.g., at 194.5 K the experimental value is $\alpha^{-1} = 3.72$ and the computed one $\alpha^{-1} = 4.22$) with the assumptions of an ideal hydrogen isotope mixture in the solid phase of palladium. It is interesting to note that after transformations (3.48) rearranges to relation (3.47) if γ is taken to have a mean value $\alpha^0 = \alpha_{HD}(4/K_{HD})^{-1/2}$.

For systems with a positive isotope effect the concentration dependence of α has been studied in the H_2–LaNi$_5$ system for the mixture H–D at temperatures of 195 and 273 K and at a pressure of 0.3 MPa ($n = 6.0$–6.5). The experimental values of the separation factor α_{H-D} obtained by the method of single adjusting for equilibrium and the computed curves [3.6, 19, 36] (solid lines) are presented in Fig. 3.14. In the calculation, values of α_{H-D} equal to 1.62 and 1.20 at 195 and 274 K, respectively, were used. In addition, at 273 K experiments to determine α_{H-D} in the H_2–SmCo$_5$ system were carried out (at $P = 0.3$ MPa the composition of the hydride phase appears to be SmCo$_5$H$_{2.5}$) [3.36]. From the experimental data also presented in Fig. 3.14 it follows that for IMC LaNi$_5$ and SmCo$_5$ the separation factors and their temperature dependences coincide. In addition, also at temperatures of 195 and 273 K and $P = 0.3$ MPa the separation factors for the H–T mixture are determined for trace amounts of tritium; the values obtained are

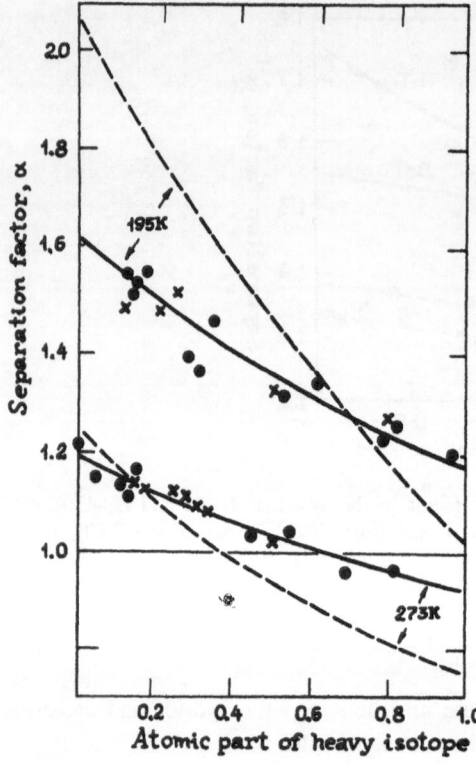

Fig. 3.14. Dependence of separation factors for the mixtures H–D (solid line) and H–T (dotted line) in H_2–$LaNi_5$ (●) and H_2–$SmCo_5$ (× systems on the composition of the gas phase)

2.10 and 1.25 at 195 and 273 K, respectively. From these experimental values of α_{HT}, the concentration dependences of α_{H-T} presented in Fig. 3.14 as dotted lines are computed.

It is evident from Figs. 3.13, 14 that, firstly, the experimental data on concentration dependence of α_{H-D} agree well with the values calculated using (3.43), secondly, for the H–T mixture the dependence is stronger than for the H–D mixture and, thirdly, the dependence of α_{A-B} on isotope composition increases with decreasing temperature. The two last conclusions are derived from the following facts: $K_{HT} < K_{HD}$ always and K_{HT} and K_{HD} decrease with decreasing temperature ($K_{HT} = 2.43$ and 1.91, $K_{HD} = 3.19$ and 2.87 at 273 and 195 K). The concentration dependence of α_{D-T} is the weakest, since the HMIE reaction $D_2 + T_2 = 2DT$ does not result in a significant deviation from the equiprobable distribution of deuterium and tritium in the molecules involved in the reaction.

In systems with a positive isotope effect, decreasing α_{A-B} with increasing heavy isotope concentration can lead to the inversion of the isotope effect. It is rather sharply defined in the H_2–$LaNi_5$ system at 273 K. The separation factors α_{H-T} and α_{H-D} decrease to 1 as the tritium or deuterium concentration increases. At a tritium concentration of over 40 at% $\alpha_{H-T} < 1$, i.e. the isotope effect becomes negative and increases with increasing tritium concentration. A similar pattern is observed for α_{H-D} at a deuterium concentration of over 60 at%.

In summary, we emphasize the importance of theoretical analysis of the concentration dependence of α and its calculation by (3.43) or (3.47) for mixtures of protium or deuterium with tritium of high concentration, because the experimental determination of α_{TH} and α_{TD} is a difficult task and these values are practically absent in the literature. Usually the literature contains only values of α_{HT} and the question of the separation factor values in the range of high tritium concentration is not touched, even in studies where isotope effects in the H_2–Pd system are considered in terms of solution of the tritium concentration problem (the sole exception is [3.37], in which temperature dependences of both α_{AB} and α_{BA} are presented for all binary mixtures of hydrogen isotopes). A different concentration dependence of α_{H-T} is observed in this particular system. It is evident from the curve presented in Fig. 3.13 that α^{-1} increases by a factor 1.7 as the tritium concentration increases. Since in a separation column repeated stages lead to multiplication of the separation effect in accordance with the Fenske equation ($K = \alpha^N$, where K is the separation degree and N is the number of theoretical stages of separation in the column), even for a comparatively moderate increase of α the achievable separation degree increases dramatically (especially for a large number of separation stages N).

3.5 Relation of Separation Factors to the Nature of Hydrides of Metals and Intermetallic Compounds

The development and performance of investigation methods such as elastic and inelastic neutron scattering, neutron diffraction and NMR, provides new information about the nature of hydrogen dissolved in metals and IMC.

The α-phase is the best studied i. e., the range of dilute solutions of hydrogen in transition metals and IMC. The nature of hydrogen dissolved in transition metals was discussed in the literature [3.38] over many years. This discussion concentrated on the question: in what form is hydrogen located in the space between sites, namely, in the form of atomic H, as H^+, or as H^-? We will not dwell long on this question, since a diversity of reviews is available [3.39–41]. Let us use the screened proton model [3.42,43] to interpret the experimental data. Take the simple qualitative model of a metal, in which the positive charge of the metal ions is distributed as a charge neutralizing the electron gas. Now consider the effect of proton injection into the hypothetical metal. Electrons group around the proton and so increase the electron density n at a distance r from the proton according to

$$\Delta n(r) = \frac{\lambda^2}{4\pi r} e^{-\lambda r}, \tag{3.49}$$

where $1/\lambda$ is the screening length, which according to *Herzfeld* and *Goeppert-Mayer* [3.42] is expressed as

$$\lambda_0^2 = \frac{4\pi q^2}{\partial\mu_e/\partial n} \tag{3.50}$$

where λ is equal to λ_0, μ_e is chemical potential of an electron in the absence of an electrostatic field and n is the concentration of electrons. It is known [3.44] that μ_e is equal to the Fermi energy. This enables one to write (3.50) as

$$\lambda_0^2 = 4\pi q^2 N_0, \tag{3.51}$$

where N_0 has the dimension of the number of electron states per unit energy and unit volume. Estimation of λ_0 for transition metals gives values from 1.5 to $7\,\text{Å}^{-1}$ [3.43], i.e., the screening charge is concentrated in a region comparable to interatomic distances.

The energy of the screened proton in the crystal lattice of the metal can be found from pseudo-potential theory [3.45] using the equation suggested in [3.43, 46]

$$\varepsilon = A\frac{e^{-\lambda r}}{r}, \tag{3.52}$$

where A is a constant involving the parameters that characterize the pseudopotential of the core of metal ions, ionic charge, and the number of adjacent metal atoms.

It follows from (3.52) that the energy of the screened proton in the crystal lattice depends on the distance Me–H and in particular on the radius $(1/r)$ of hydrogen in the space between sites. For minor changes Δr in the distance between hydrogen and metal atoms one can derive

$$\frac{\partial^2\varepsilon}{\partial[\Delta r]^2} = \frac{2A\lambda}{r}\exp(-\lambda r). \tag{3.53}$$

Thus the frequency of the local modes of hydrogen atoms of mass m is expressed as

$$\omega^2 = \frac{\partial^2\varepsilon/\partial[\Delta r]^2}{4\pi^2 m} = \frac{2A\lambda}{4\pi^2 mr}\exp(-\lambda r). \tag{3.54}$$

When replacing protium with deuterium, the values A, λ, and r change slightly and thus, the known expression can be derived

$$\frac{\omega_{H_2}}{\omega_{D_2}} = \frac{m_D}{m_H} = 2. \tag{3.55}$$

Let us analyze the possibility of applying the screened proton model to real hydrogen–transition metal systems. As noted in Chap. 2, calculations of the phase diagrams using this model agree well with the experimental data for the range of both α- and β-phase existence in view of additional terms representing the H–H interaction.

By means of (3.54) *Ebisuzaki* [3.43] determined the frequency of hydrogen atom local modes in Ni using the data for ω_H in Pd, and revealed good correlation with the experimental data ($\omega_H^{Ni}/\omega_H^{Pd} = 2$ compared with experimental value $\omega_H^{Ni}/\omega_H^{Pd} = 2.1$). It was assumed that Ni and Pd have equal values of the constants A and λ. As noted above, the constant A involves quantities reflecting the nature of the metal and the interstitial position of hydrogen. Hence, comparison of ω_H for different metals can be performed without determination of the constant A provided the metals have equal structure, are in the same group of the periodic system, and hydrogen occupies the same interstices in the crystal lattice.

Table 3.3 presents the results of the determination of frequencies of hydrogen atom local modes in the crystal lattice of a number of transition metals, and the values of the distance Me–H found experimentally both for solid solutions of hydrogen and for hydride phases.

As is known, the process of hydride formation can be followed by one of the following structure modifications: (1) change of the metallic lattice parameters; (2) deformation of the metallic lattice (BCC-BCT, FCC-FCT); or (3) transformation of crystallographic structures (HCP-BCC-FCC).

In the first case slight differences in the value of ω_H are expected; in the second case, an additional nondegenerate frequency arises in the spectra of neutron scattering due to the deformation of the lattice along the z-axis.

It is to be noted that deformation of the metallic lattice can modify the hydrogen positions, e. g. it is possible for hydrogen atoms to move from tetrahedral into octahedral interstices and conversely. The first two cases are most often observed upon formation of hydrides by metals in "soft" conditions. The third case is typical for metals forming hydrides at pressures of 2000 bar and above (Cr, Mn, Fe, Co).

In the majority of hydride-forming transition metals, apart from palladium, hydrogen occupies tetrahedral interstices. This significantly simplifies the comparison of different hydrides to one another.

As noted above, the relation (3.52) enables us to analyze, which parameters effect the energy of hydrogen atoms and, hence, the separation factor of hydrogen isotopes in hydrogen–metal/IMC-hydride systems.

3.5.1 Influence of the Geometrical Factor on α in Hydrogen–Metal-Hydride Systems

Many investigations of hydrides of metals and IMC have shown that the geometrical factor (distance Me–Me and H–H) is one of the basic factors determining the possibility of hydride formation and the structure of alloys.

Comprehensive analysis and calculations of Me–H distances have been performed by *Pous* and co-workers [3.66, 67]. Let us analyze these data applied to the dependence $\alpha(\varepsilon)$ on the nature of metallic hydrides. On the basis of (3.52) it can be assumed that the greater the distance Me–H, the lower the energy and, hence, the separation factor. This is true provided $A = \text{const}$ and $\lambda = \text{const}$.

The value $1/\lambda$ represents the space occupied by a hydrogen atom in the crystal lattice.

Table 3.3. Frequencies of local modes of hydrogen atoms in the crystal lattice of metals

MeH(D)	$\hbar\omega_{H(D),i}$ [meV]			r_{Me-H} [Å]	Reference
	$i = 1$	$i = 2$	$i = 3$		
$TaH_{0.15}$	120	170	–	1.65	[3.47]
$TaD_{0.22}$	$119/\sqrt{2}$	$167/\sqrt{2}$	–	1.65	[3.48]
$TaH_{0.7}$	130	180	–	1.65	[3.49]
$NbH_{0.05}$	110	180	–	1.65	[3.50]
$NbD_{0.6}$	$113/\sqrt{2}$	$158/\sqrt{2}$	–	1.65	[3.51]
$NbD_{0.75}$	$120/\sqrt{2}$	$170/\sqrt{2}$	–	1.65	[3.52]
$NbD_{0.95}$	120	170	–	1.65	[3.53]
NbH_2	148	–	–	1.98	[3.54]
$VH_{0.04}$	120	170	–	1.51	[3.50]
V_2H	56	130	230	1.65	[3.56]
V_2D	39	87	164	1.65	[3.56]
$VH_{1.5-1.7}$	160	–	–	1.85	[3.57]
$VH_{1.92}$	165	–	–	1.85	[3.54]
$TiH_{0.05}$	141	–	–	1.803	[3.55]
$TiH_{0.14}$	120	141	171	1.803	[3.55]
TiH_2	149	–	–	1.93	[3.55]
$TiH_{0.07}$	105.5	162	–	1.803	[3.58]
$TiD_{0.09}$	108.5	–	–	1.803	[3.58]
$ZrH_{0.03}$	143.1	–	–	1.973	[3.58]
$ZrD_{0.05}$	105.0	–	–	1.973	[3.58]
$ZrH_{0.05}$	144	–	–	1.973	[3.55]
ZrH_2	141	–	–	2.04	[3.54]
$PdH_{0.01}$	68.5	–	–	1.94	[3.59]
$PdH_{0.6}$	54	–	–	2.04	[3.55]
$LaH_{1.94}$	103	–	–	2.45	[3.60]
$PrH_{1.94-2.0}$	108	–	–	2.39	[3.60]
$HoH_{1.98}$	126	–	–	2.24	[3.54]
ErH_2	128	–	–	2.22	[3.54]
YbH_2	130	–	–	2.30	[3.61]
CaH_2	125	–	–	2.32	[3.61]
SrH_2	116	–	–	2.49	[3.61]
BaH_2	100	–	–	2.67	[3.60]
$CeH_{0.5}$	96.2	–	–	2.23	[3.62]
$CeH_{1.98}$	106	–	–	2.42	[3.63]
YH_2	127	–	–	2.25	[3.64]
$UH_{2.67}$	112	–	–	2.32	[3.65]

Detailed analysis of the volume occupied by a hydrogen atom in the lattice of hydrides of metals and IMC were performed by *Fukai* [3.68–70], who, in particular, corroborated the so-called "rule of three" suggested by *Baranowski* [3.71]. Analysis of the published data has shown that for solid solutions of hydrogen in FCC (Ni, Pd), HCP (Fe, Y), and BCC (V, Nb, Ta) metals and for bihydrides of the fluorite type (TiH$_2$, ZrH$_2$, YH$_2$, NbH$_2$, TaH$_2$, ScH$_2$) expansion of the lattice volume per single hydrogen atom falls in the narrow range ($\Delta V_H = 2.8 \pm 0.2\,\text{Å}^3$). This corresponds to the hydrogen radius in the crystal lattice 0.87 Å. It should be noted that the radius of a hydrogen atom with zero valency is expressed as:

$$r_{H^0} = 0.5(r_{H^-} + r_{H^+}) = 0.90\,\text{Å}. \tag{3.56}$$

These data enable one to use (3.52) qualitatively when considering hydrides, since the value $\lambda = 1/r_H$ practically does not change in the transition from solid solutions to hydrides and one can also consider this value as a constant when comparing the above-mentioned metals.

Thus, for $\lambda = $ const, to estimate $\alpha(\varepsilon)$ it is necessary, in addition to r_{Me-H}, to know the value of the constant A. Experimental determination of this constant is significantly more difficult than the direct determination of the separation factor.

Let us use the empirical dependence [3.54] obtained for bihydrides of rare-earth elements to assemble the experimental data on ω_H obtained by the INSS method and the experimental data on determination of separation factors.

Figure 3.15 presents the empirical dependence $\varepsilon = 410\,\text{meV Å}^{3/2} \times r^{-3/2}$ and the experimental values for solid solutions and hydrides of some transition metals. The error in determination of ε varies from ± 1 to $\pm 4\,\text{meV}$ depending on the reference.

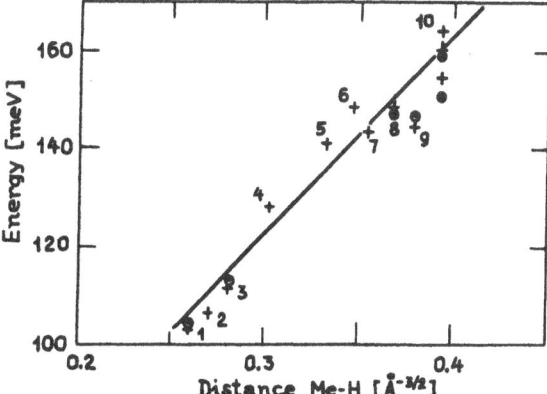

Fig. 3.15. Dependence of the local mode energy of hydrogen atoms on distance Me–H. (1) La, (2) Ce, (3) U, (4) Er, (5) β-Zr, (6) β-Ti, (7) α-Zr, (8) Nb, (9) Ta, (10) V. + experimental data; ○ data calculated from values of α

Values of r_{Me-H} are experimentally obtained by the neutron diffraction method; the calculated ones are taken from [3.66]. As is shown in Sect. 3.3, the separation factor for any pair of hydrogen isotopes can be computed using the harmonic-oscillator model. For some metals, such as V, Zr, Ta, Nb, Ti, the experimental data on separation factors for the mixtures H–D and H–T are available, for other metals this value is computed from the isotherms of sorption of protium and deuterium or tritium. The values of ε calculated from the α values are also presented in Fig. 3.15. It is found from Fig. 3.15 that quantitative estimation of α is possible, provided the hydrogen isotopes are located in tetrahedral interstices, by means of the empirical dependence and the known values of r_{Me-H}.

Only in palladium and vanadium (V_2H) (at definite hydrogen concentrations) does hydrogen occupy octahedral interstices. Many studies have been devoted to the investigation of these systems and recent attention has focused on the vanadium–hydrogen system. This system is of interest since hydrogen occupies both tetrahedral and octahedral positions, which should affect the value of the isotope effect.

The experimental data for the separation factors in hydrogen isotope–vanadium systems for the range of low concentrations of the heavy isotope [3.9, 72] and the data of the present authors on determination of α_{H-D}^0 by the method of laser desorption are summarized in Tables 3.4, 5.

Table 3.4. Temperature dependence of separation factors for the isotopes H–T

| VH$_x$ | Method of single adjusting for equilibrium | | | Chromatographic method [3.72] | |
	T [K]	α_{HT}	Reference	T [K]	α_{HT}
VH$_{0.65}$	483	1.1	[3.9]	676	1.07*
VH$_{0.85}$	303.6	1.4	[3.9]	523	1.05
VH$_{0.7}$	519	1.11	[3.9]	473	1.12*
VH$_{0.7}$	298	1.17	++	433	1.12*
VH$_{0.7}$	298	0.75+	++	373	1.16*
VH$_{0.7}$	373	1.05	[3.72]	333	1.19*
	373	1.08	[3.72]	313	1.26
	373	1.11	[3.72]		
	373	1.13	[3.72]		
	373	1.18	[3.72]		
VH$_2$	318.2	1.61	[3.9]		
	301.2	1.73*	[3.9]		
	273	1.87*	[3.9]		

* average value for two or more experiments
+ fast sample cooling   ++ data of this work

The works [3.73–75] are devoted to the study of the phase diagrams of the systems V–H$_2$, V–D$_2$, V–T$_2$ in the hydrogen concentration range $n \leq 1$. All three phase diagrams [3.75] are presented in Fig. 3.16 for comparison.

Table 3.5. Temperature dependence of separation factors for V-hydride and the isotopes H–D

T [K]	Single adjusting for equilibrium (LR), α_{HD}	LAMMA, α^0_{H-D}
473	1.09	1.09
373	0.83	0.83
298	0.50	0.47
297	0.47	0.47
298	1.08[+]	0.99[+]
298	–	1.08[+]

[+] fast sample cooling

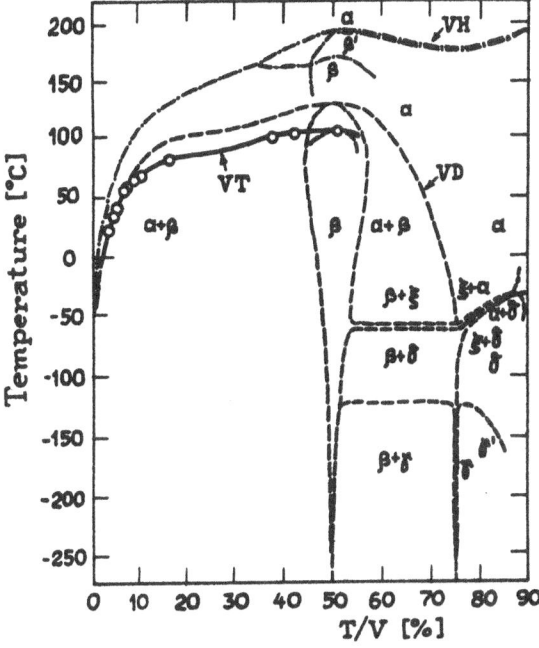

Fig. 3.16. Phase diagrams in hydrogen isotopes–vanadium systems [3.75]

As is seen, the phase diagram of the V–H$_2$ system differs essentially from those for the systems with a heavy isotope at concentrations $n > 0.5$. Authors [3.76] believe that this has to result in a considerable isotope effect. A number of works are devoted to a thorough study of this question. We refer to those of them, which, in our opinion, are of the most interest, since they contain the neutron scattering data over a wide temperature range and include the temperatures of phase transformations of the first type ($\alpha \rightarrow \beta$) and the second type ($\beta \rightarrow \beta'$) [3.56, 77, 78]. It is known that in the α-phase both protium and the heavy isotope are located in tetrahedral interstices. The available data on the frequencies of optical modes [3.79] show that doubly degenerate high-energy frequencies ω_H = 180 meV and ω_H = 113 meV arise due to the (BBC-BCT) lattice deformation. At the $\alpha \rightarrow \beta$ transition hydrogen occupies low-energy (55 meV) [3.77] octahedral interstices, whereas deuterium atoms, for instance, can remain in tetrahedrons of the α-phase, which leads to a significant isotope effect in the temperature range 313–453 K at $n = 0.7$. In [3.72] the separation factors α_{HT} are found at $n = 0.7$ by the chromatographic method in just this temperature range.

As it is evident from Table 3.4, no essential change of α_{HT} is observed with changing temperature. It is to be noted that the data obtained by the method of single adjusting for equilibrium [3.9, 72] diverge considerably in the range of low temperatures (313–373 K). Because of the absence of any similar investigations for the mixture protium–deuterium, we carried out a study of the isotope effect by the new method of laser desorption (LAMMA or laser microprobe mass analysis) and by the routine method of single adjusting for equilibrium (LR or loop reactor).

To find out the influence of phase transformations on α a sample of vanadium (V) evacuated at $T = 570$ K to $P = 1.3 \times 10^{-6}$ mbar is exposed to a protium-deuterium mixture (50 : 50 LAMMA, 93 : 7 LR) at $P = 0.1$ MPa. Sample cooling to the experimental temperature is performed in two ways: slow (0.2 K/min) and fast (10 K/min). One takes into account, on the basis of data in [3.9], that phase and isotope equilibrium of hydrogen with vanadium is attained in both cases. For slow cooling protium and deuterium enter isomorphous β_H and β_D phases and occupy octahedral interstices; for fast cooling a significant portion of the deuterium atoms can remain in tetrahedral positions. The first case leads to the inversion of the isotope effect, the second, to its considerable reduction. In the transition range (nonequilibrium in terms of phase transformations of the first type) the effective value of $\overline{\alpha}$, which includes the possibility of simultaneous occupation of octahedral and tetrahedral positions by both isotopes, is given by

$$\overline{\alpha} = \varphi \alpha_{tet} + (1 - \varphi)\alpha_{oct}, \tag{3.57}$$

where φ is the fraction of atoms in tetrahedral positions. Equation (3.57) is valid on condition that the occupation probabilities for protium and deuterium atoms are equal.

Figure 3.17 shows the straight lines plotted in coordinates $\ln \alpha$–$1/T$, which are calculated by the harmonic-oscillator model for the case when both isotopes are located in tetrahedral interstices or in octahedral ones. The experimental data summarized in Tables 3.4, 5 are also presented in this figure. As is evident from Fig. 3.17, at high temperatures the values fall on lines 1 and 2; at low temperatures, on lines 2 and 3. In the transition temperature range (fast cooling) the experimental data are located between lines 2 and 3 for α_{H-D} and lines 1 and 4 for α_{H-T}, so no abrupt change of the isotope effect is observed.

The question of phase transformations of the second type is still being discussed in the literature. It can be noted that, although the anharmonic nature of this system shows itself markedly [3.56], the consideration of the isotope effect shows that phase transitions of the second type ($\beta - \beta'$) do not strongly affect α in the hydrogen–vanadium system.

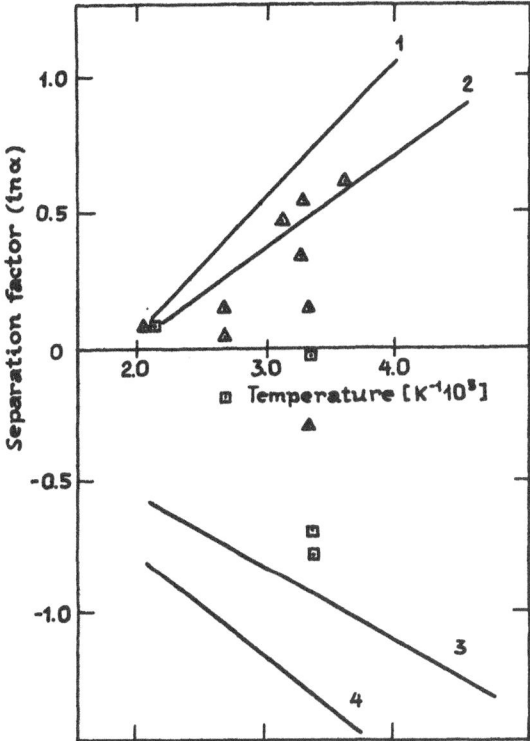

Fig. 3.17. Dependence of separation factors on temperature in the hydrogen–vanadium hydride system. (*1*) α_{H-T}^0, (*2*) α_{H-D}^0 (α-phase); (*3*) α_{H-D}^0, (*4*) α_{H-T}^0 (β-phase); $\square, \triangle$ experimental values of α_{H-D}^0 and α_{H-T}^0, respectively

In terms of hydrogen isotope separation, vanadium bihydride is the most interesting object. Both γVH_2 and γVD_2 have the same FCC lattice of the CaF_2 type, in which hydrogen atoms occupy equivalent tetrahedral positions. The available data on the energy of modes $\varepsilon = 160\,meV$ [3.54, 57] enable one to compute the temperature dependence of α.

As shown above, equation (3.36) allows averaged values of the frequencies to be used for the calculation of the β-factor and hence of α. Since the value 160 meV is close to the average value of the doubly degenerate modes in αVH (157.7 meV), the straight lines in Fig. 3.17 coincide for the α- and γ-phases. The experimental data for the hydrogen–γVH_2 system agree well with the computed results.

Thus the harmonic oscillator model, in spite of a number of assumptions, can be successfully applied to evaluable separation factors and their temperature dependence in real systems.

In contrast to metals in IMC two geometrical parameters must be considered: r_A/r_B defines the stability of the IMC being formed and r_{IMC-H} is the average distance between hydrogen and atoms of elements A and B.

A comprehensive analysis of the influence of the ratio r_A/r_B on IMC stability is presented in [3.80, 81]. Most common IMC related to the Laves phases are formed at a ratio r_A/r_B of 1.05–1.68, the ratio for close packed FCC (MgCu$_2$) and HCP (MgZn$_2$) structures is equal to 1.225 and corresponds to ideal packing of spheres, i.e. to the undeformed crystal lattice. For compounds RMe$_5$ having CaCu$_5$ HCP r_A/r_B changes from 1.29 to 1.61.

The interatomic distance r_{IMC-H} is related to the values of r_A and r_B in accordance with the Pauling equation [3.82]:

$$r_{A-H} = r^0_{A-H} - 0.3 \lg \left(\frac{\varepsilon_A x}{cN_H} \right) - 0.3 \lg \left(\frac{1}{cN_A} \right),$$

$$r_{B-H} = r^0_{B-H} - 0.3 \lg \left(\frac{\varepsilon_B x}{cN_H} \right) - 0.3 \lg \left(\frac{1}{cN_B} \right), \tag{3.58}$$

where εx is the number of valence electrons of the metal involved in the Me–H bond formation; x is the atomic composition of MeH$_x$; $r^0_{Me-H} = 0.5(r^0_{Me-Me} + D^0_{H-H})$, $D^0_{H-H} = 0.64$ Å is the length of covalent bond; cN_H is the number of hydrogen atoms surrounding an atom of the metal; cN_{Me} is the number of metal atoms surrounding a hydrogen atom.

In the majority of IMC hydrides, hydrogen primarily occupies tetrahedral interstices A$_2$B$_2$ and the value of r_{Me-H} can be taken as the average, i.e.

$$r_{IMC-H} = 0.5(r_{A-H} + r_{B-H}). \tag{3.59}$$

However, the calculation using [3.58] is possible only if $n_H/n_{IMC} > 1$; otherwise, it appears to be impossible to determine r_{B-H}. Thus it is convenient to use the distance between metals as a geometric parameter for comparison of one IMC to another. The distance can be found from the equation.

$$r_{Me-Me} = r^0_{Me-Me} - 0.6 \lg \left(\frac{n_e}{cN_{Me}} \right) \tag{3.60}$$

where n_e is the number of valence electrons involved in the metallic bond, cN_{Me} is the coordination number in the metallic lattice. For IMC let us take this value as the average for atoms A and B:

$$r_{IMC} = 0.5(r^0_{A-A} + r^0_{B-B}) - 0.3 \lg \left(\frac{n_{eA}}{cN_A} \right) - 0.3 \lg \left(\frac{n_{eB}}{cN_B} \right). \tag{3.61}$$

Table 3.6 shows the values of the parameter r_{IMC}, which are averages of the computed and the experimental ones, for a number of IMC [3.67].

Based upon the above approach, one would expect the largest separation factors for TiFe, TiCr and TiCr$_2$, since r_{IMC} for these compounds is smaller than for other IMC.

Table 3.6. Interatomic distances in IMC

IMC	Lattice type	r_{IMC} [Å]	Critical parameter r_{IMC} [Å]
TiCr	BCC + BCT	2.70	2.58–2.60
TiFe	BCC + BCT	2.60	2.58–2.60
TiCo	BCC + BCT	2.62	2.58–2.60
TiV	BCC + BCT	2.73	2.58–2.60
TiNi	BCC + BCT	2.61	2.58–2.60
TiMo	BCC + BCT	2.76	2.58–2.60
TiZr	BCC + BCT	2.95	2.58–2.60
ZrCo	BCC + BCT	2.78	2.58–2.60
$TiCr_2$	BCC + BCT	2.8	2.87
ZrV_2	FCC	2.99	2.87
$ZrCr_2$	FCC	2.89	2.87
$ZrFe_2$	FCC	2.83	2.87
$ZrCo_2$	FCC	2.79	2.87
HfV_2	FCC	2.98	2.87
$TiCr_2$	HCP	2.70	3.10
ZrV_2	HCP	2.92	3.10
$ZrCr_2$	HCP	2.81	3.10
$ZrMn_2$	HCP	2.80	3.10

3.5.2 Dependence of the Separation Factor on the Electronic Structure of Metals and IMC

As is obvious from (3.61), the geometric parameter r_{IMC} for metals or its average value for IMC decreases with increasing n'_e. This leads to a rise in the energy ε and, hence α. This approach is valid if hydrogen is localized in a single type of interstitial site.

In the first series of transition metals, the distance Me–Me decreases and the bond energy increases up to Cr and thereafter remains constant. It can lead to a similar dependence of α, however, the probability that octahedral (low-energy) interstices are occupied increases with increasing d-character; the effective separation factor must then fall and appears to be the lowest for Pd, in which hydrogen atoms occupy only octahedral interstices.

A similar pattern is also to be observed for IMC. The quantitative analysis of this question is a difficult crystallochemical problem.

Table 3.7 presents the available experimental values of separation factors α_{HT} and α_{HD} for hydrogen–metals-hydrides and IMC systems.

As is evident from Table 3.7, the majority of experimental values are obtained for the H–T mixture in the range of low tritium concentrations; for the H–D mixture the majority of the data are obtained in [3.91] by the chromatographic method at the relatively high temperature of 333 K. However, in this method α can

Table 3.7. Experimental values of separation factors for hydrogen isotopes

Hydrides of Me and IMC	T [K]	α_{HT}	α_{H-D}	References
TiH_2	623	0.67	–	[3.83]
TiH_2	373	0.45	–	[3.84]
VH_2	273	1.91	1.73 (313* K)	[3.9]
$CrH_{0.6}$	423	–	0.97*	[3.85]
$NiH_{0.5}$	298	–	0.92*	[3.85]
ZrH_2	673	1.07	–	[3.84]
NbH_2	333.6	–	1.73	[3.9]
$PdH_{0.4}$	273	0.34	0.46	[3.12]
UH_3	600	–	0.75	[3.10]
$TiVH_{4.15}$	313	1.18	–	[3.83]
$TiCrH_{2.35}$	313	1.54	–	[3.83]
$TiMnH_{1.99}$	313	1.37	–	[3.83]
$TiFeH_{1.88}$	273	0.92	–	[3.83]
$TiCoH_{1.44}$	313	0.85	–	[3.83]
$TiNiH_{1.44}$	313	0.74	–	[3.83]
$TiFe_{0.6}Mn_{0.2}H_{1.67}$	313	1.00	–	[3.83]
$TiMoH_{2.99}$	313	1.61	–	[3.86]
$ZrCoH_{0.8-1.5}$	473	–	1.14*	[3.87]
$ZrNiH_3$	300.6	1.05	–	[3.83]
$TiCr_2H_{1.48}$	273	2.03	–	[3.83]
$TiCr_2H_{1.68}$	253	2.01	–	[3.83]
$TiMn_{1.5}H_{2.5}$	195	3.00	1.6 (228* K)	[3.7]
$TiMo_2H_{1.1}$	253	1.87	–	[3.86]
$TiCrMnH_{1.28}$	273	1.80	–	[3.83]
$TiCrMnH_{2.19}$	253	2.05	–	[3.83]
$TiCrMnH_{3.0}$	195	3.25	1.7 (228* K)	[3.8]
$Ti_{0.8}Zr_{0.2}Cr_{1.8}H_{2.8}$	195	3.10	–	[3.8]
$Ti_{0.8}Zr_{0.2}CrMnH_{2.9}$	195	3.30	–	[3.8]
$TiMn_{1.4}Ni_{0.1}H_{2.3}$	195	2.80	–	[3.8]
$ZrV_2H_{4.12}$	273	1.77***	–	
$ZrCr_2H_3$	298	1.90	1.60	[3.88]
$ZrMn_2H_{0.85}$	300	–	1.38	[3.89]
$ZrMn_2H_3$	273	1.75***	1.47***	
$Zr_{0.8}Ti_{0.2}Mn_2H_{2.8}$	333	–	1.10	[3.90]
$LaNi_5H_{6.6}$	195	2.10	1.62	[3.6]
$MmNi_5H_{6.6}$	273	1.29	–	[3.83]

Table 3.7 (continued)

Hydrides of Me and IMC	T [K]	α_{HT}	α_{H-D}	References
$LaCo_5H_{3.4}$	333	–	1.3**	[3.91]
$MmCo_5H_{2.8}$	333	–	1.2**	[3.91]
$SmCo_5H_3$	195	2.04	–	[3.6]
$CaNi_5H_{5.27}$	333	–	1.1**	[3.91]
$CaNi_5H_{5.5}$	298	–	1.0*	[3.92]
$LaNi_{4.5}Al_{0.5}H_{5.5}$	333	–	1.2**	[3.91]
$LaNi_4AlH_{3.8}$	333	–	1.3**	[3.91]
$LaNi_4CuH_{5.5}$	195	2.04	–	[3.36]
$LaNi_4CrH_{4.8}$	195	2.04	–	[3.36]
$LaNi_4CoH_{5.8}$	333	–	1.1**	[3.91]
$LaNi_{3.5}Co_{1.5}H_x$	333	–	1.0**	[3.91]
$LaNi_3Co_2H_{5.0}$	333	–	1.0**	[3.91]
$LaNi_3Cu_2H_{5.2}$	195	2.04	–	[3.36]
$LaNi_{2.4}Co_{2.6}H_{5.4}$	333	–	1.1**	[3.91]
$LaNi_2Co_3H_{4.8}$	333	–	1.1**	[3.91]
$LaNiCu_4H_{3.5}$	333	–	1.3**	[3.91]
$LaNiCo_4H_{4.3}$	333	–	1.2**	[3.91]
$LaNi_{0.5}Co_{4.5}H_{3.9}$	333	–	1.3**	[3.91]
$CaNi_4AlH_{1.6}$	333	–	1.4**	[3.91]
$CaNi_4CrH_{1.2}$	333	–	1.2**	[3.91]
$CaNi_4TiH_{0.7}$	333	–	1.0**	[3.91]
$CaNi_4MnH_{2.1}$	333	–	1.1**	[3.91]
$CaNi_4CoH_{1.8}$	333	–	1.2**	[3.91]
$CaNi_4CuH_{5.1}$	333	–	1.2**	[3.91]
Mg_2NiH_4	524	0.48	–	[3.84]

* data obtained from the isotherms of protium and deuterium sorption
** data obtained by chromatographic method   *** data of the present work

be determined only to a lower accuracy than in the method of single adjusting for equilibrium. Comparison of the experimental results to one another is complicated by the fact that they are obtained at different temperatures and for the H–D mixture at different concentrations of the heavy isotope. Using the harmonic-oscillator model and the concentration dependence of α (Sects. 3.3, 4) the experimental data are recalculated for temperatures of 173 and 273 K, which are the most interesting in terms of isotope separation, to facilitate their comparison.

In the work [3.83] based on data obtained for hydrides of metals and IMC of the elements of the third period, the authors have attempted to relate the value α_{HT} to the number of valence electrons per atom of metal $\alpha_{HT} = f(e/m)$. However, this comparison is performed with values of α_{HT} obtained at different temperatures. Figure 3.18 shows the dependences $\alpha_{HT} = f(e/m)$ at 173 and 273 K for metals and IMC of the third period and the dependence $\Delta H_H^{\alpha-\beta} = f(e/m)$. As is evident, these dependences have the reverse charcter and have extrema in the same range of e/m values.

A similar dependence is observed for compounds based on Zr and elements of the third period. The experimental values obtained recently by the authors for α_{HT} and α_{HD} by the method of single adjusting for equilibrium are summarized in Table 3.8.

Using both geometric and electronic factors, one can analyze the data of Tables 3.7, 8 and Fig. 3.18.

The values of α increase as the number of electrons per IMC atom increases from 4 to 5. Thereafter a maximum is observed at a value $e/m = 5.2$, the further rise of e/m is accompanied by a decrease of α. IMC with more than 6 electrons have low values of α. Compounds of Zr, Ti, and La with transition elements Ni, Co, Fe fall into this group of IMC. In these compounds, along with tetrahedral positions, hydrogen also occupies octahedral sites [3.93–96], leading to a decrease of the experimental value of $\bar{\alpha}$. In the case of isoelectronic IMC and metals, the geometric factor becomes crucial: the data for ZrCr$_2$ and ZrVMn illustrate this fact. The insignificant influence of the substituent metal in compounds RNi$_4$Me is related to the fact that this metal does not essentially affect the ratio n_{tet}/n_{oct} and, hence, these IMC have similar values of α.

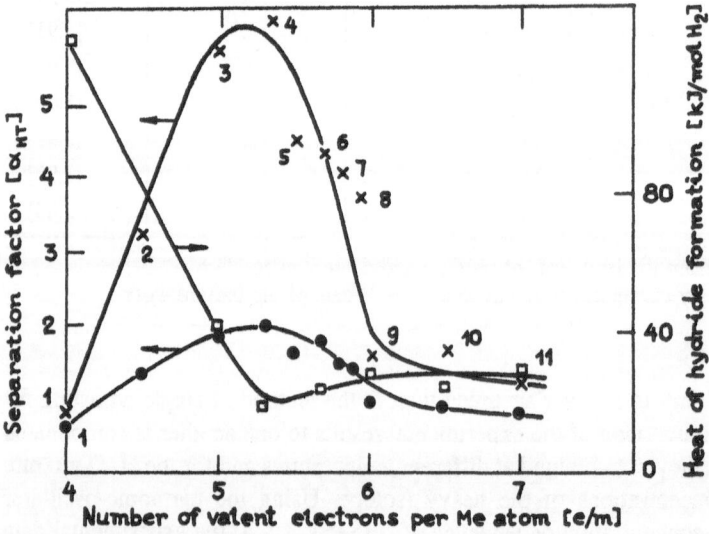

Fig. 3.18. Dependence of separation factors α_{HT} and the heat of hydride formation $\Delta H_H^{\alpha-\beta}$ on the number of valence electrons e/m per atom of metal or IMC. □ $\Delta H_H^{\alpha-\beta}$; × α_{HT} at 173 K; • α_{HT} at 273 K; (1) Ti, (2) TiV, (3) TiCr, (4) TiCr$_2$, (5) TiMn, (6) TiCrMn, (7) TiMn$_{1.5}$, (8) TiMn$_{1.4}$Ni$_{0.1}$, (9) TiFe, (10) TiCo, (11) TiNi

Table 3.8. Experimental data on simultaneous determination of separation factors

IMC	T [K]	α_{HT}	α_{HD}	IMC	T [K]	α_{HT}	α_{HD}
ZrV_2	250	2.01	–	$ZrMn_{3.8}$	240	1.76	–
	273	1.77	–		250	1.69	1.41
	298	1.52	–		273	1.51	–
	373	1.20	–		294	1.39	1.19
					323	1.27	–
$ZrCr_2$	273	2.32	–				
	298	1.90	1.60	$ZrMn_2Cr_{0.8}$	250	2.04	1.69
	323	1.55	1.37		273	1.70	1.45
	353	1.35	1.07		297	1.54	1.33
	373	1.19	–		323	1.36	1.23
$ZrMn_2$	239	2.08	1.66	$ZrMnV$	250	1.79	–
	273	1.75	1.47		273	1.65	–
	296	1.55	1.27		298	1.44	–
	323	1.30	1.17		323	1.19	–
$ZrMn_{2.8}$	239	1.82	–				
	250	1.72	1.41				
	273	1.54	–				
	294	1.41	1.21				
	323	1.26	1.10				

Comparison among IMC shows that the largest values of α can be found for compounds of the Laves phases AB_2 with FCC and FCT structure. For compounds of type AB, AB_3, A_2B_7 and AB_5, α is lower because of the greater probability of occupation of octahedral sites as is evident from Tables 3.7, 8.

The analysis of interatomic distances and the least sizes of tetrahedral interstices performed in [3.97] showed that IMC hydrides form and remain stable at room temperature provided r_{IMC} exceeds 2.65, 2.75 and 2.80 Å for BCC, HCP and FCC, respectively. Using these parameters and average values of e/m it appears to be possible to choose the optimum IMC composition with three and more components. The chemical composition can be selected so as to optimize the following parameters: high hydrogen capacity, suitable thermal stability, and large value of separation factors.

For example, substitution of a transition metal by an isoelectronic metal having larger atoms results in an increase of stability and a decrease of the separation factor. By combining different transition metals with one another, it is possible to synthesize the required hydrides.

The most interesting data can be obtained for the IMC $Cr_2(Zr, Ti)$, $(CrMn)_2(ZrTi)$, $(CrMn)_2Ti$, $(CrV)_2Zr$, $(CrMn)_2Zr$.

Hyperstoichiometric compounds $ZrMn_2Mn_x$ are included in a separate group. Table 3.8 presents the data on determination of separation factors in these systems.

It is known [3.98] that in the series $ZrMn_2Mn_x$ manganese holds the zirconium positions in the crystal lattice. This results in a decrease of the lattice parameters and in a larger number of tetrahedral interstices AB_3 of smaller size.

The probability of the interstices AB_3 being occupied increases with increasing Mn concentration [3.99] and can lead to a rise in $\overline{\alpha}$. On the other hand, an increase of e/m causes $\overline{\alpha}$ to decrease.

It is obvious from Table 3.8 that in hydrogen–IMC–hydride systems $\overline{\alpha}$ for $ZrMn_2$ is higher than for $ZrMn_{2.8}$. For the latter, the value is approximately equal to that for $ZrMn_{3.8}$, i.e. the electronic factor in hyperstoichiometric compounds is compensated by the geometrical factor.

For the compounds $ZrMn_2Me_{0.8}$ (Me=Fe,Co,Ni) and $ZrCr_2Me_{0.8}$ a similar pattern is observed [3.100, 101]. It is to be noted that these compounds are unstable and prone to fast hydrogenolysis followed by release of Cr due to the narrow range of homogeneity for $ZrCr_{2+x}$ ($-0.5 \le x \le 0.2$), which is why the compound $ZrCrFeCo_{0.8}$ is of considerable interest [3.102].

The available data on changes of the crystal lattice volume and values of $\Delta H_H^{\alpha-\beta}$ lead one to expect a high value of α for the compound $ZrCrFeCo_{0.8}$. However, high pressures of hydride formation and strong dependence of the pressure on capacity render this compound less useful in terms of isotope separation.

In summary, it should be noted that simple criteria for preliminary selection of IMC and their directed synthesis are important in practice.

The values r_A/r_B and e/m can be applied as such criteria for the Laves phases. They represent both electronic and geometric factors.

Figure 3.19 shows the range of r_A/r_B and e/m into which fall practically all IMC of interest in isotope separation in terms of thermodynamic considerations.

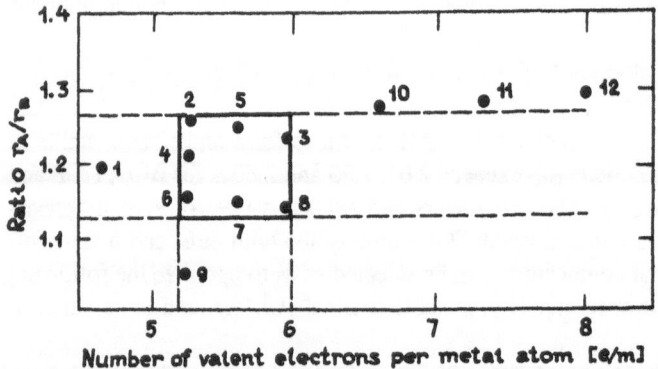

Fig. 3.19. Allowable changes of geometric and electronic factors for tentative choice of IMC with a high thermodynamic isotope effect of hydrogen with their hydride phases: (1) ZrV_2, (2) $ZrCr_2$, (3) $ZrMn_2$, (4) $ZrVMn$, (5) $ZrCrMn$, (6) $TiCr_2$, (7) $TiCrMn$, (8) $TiMn_{1.5}$, (9) $TiMo_2$, (10) $ZrFe_2$, (11) $ZrCo_2$, (12) $ZrNi_2$

3.6 Dependence of Separation Factors on Temperature and Pressure

The quantum-statistical method of calculating the thermodynamic isotope effect outlined in Sect. 3.3 describes well the temperature dependence of the separation factors in hydrogen-isotope–metal- and IMC–hydride systems [3.6–8, 84, 88].

The majority of available data relate to trace amounts of tritium and low deuterium concentrations, corresponding to (3.18, 28). For the chemical isotope exchange reaction (3.18) the equation for the separation factor can be written with regard to (3.23) in the following form:

$$\alpha_{AB} = \frac{K}{K_{\infty}}$$
$$= \frac{u_{A_2}}{u_{AB}} \frac{[1 - \exp(-u_A)]^3/[1 - \exp(-u_B)]^3}{[1 - \exp(-u_{A_2})]/[1 - \exp(-u_{AB})]} \frac{\exp[0.5(u_A - u_B)]^3}{\exp[0.5(u_{A_2} - u_{AB})]} . \quad (3.62)$$

The first factor is constant and the other two depend on temperature. The temperature dependence of the second factor should be considered only at elevated temperatures; at low temperatures it is equal to 1. Thus the temperature dependence of α_{AB} is determined mainly by the third factor reflecting the influence of the change of zero-point energy $\Delta\varepsilon^0$ of hydrogen molecules and atoms on α_{AB} during the isotope exchange reaction. This allows one to simplify the dependence of α_{AB} on temperature:

$$\ln \alpha_{AB} \approx a + b/T , \quad (3.63)$$

where a and b are constants, which in turn can be derived from the known relation for equilibrium constants, $\Delta G = -RT \ln K$, for the isotope exchange reaction

$$-RT \ln K = \Delta H_{(AB)} - T \Delta S_{(AB)} \quad (3.64)$$

or

$$\ln \alpha_{AB} = \ln 2 + \frac{\Delta S_{(AB)}}{R} - \frac{\Delta H_{(AB)}}{RT} . \quad (3.65)$$

So, if the experimental dependence is adequately described by (3.63), it is possible to find the values of the thermodynamic parameters of the isotope exchange reaction. Table 3.9 presents the values of the constants a and b for a number of metals and IMC.

It is to be noted that in all cases, apart from Pd, appreciable deviation of the experimental data from those computed by the harmonic-oscillator model is observed at $T > 323$ K. For Pd the authors took into account, firstly, the anharmonicity using the relations $\omega_H = 1.5\omega_D = 1.87\omega_T$ and, secondly, as shown in [3.21] that at low ω_H the calculated temperature dependence has an appreciably larger range of linear behaviour (up to 500 K) in the coordinates $\ln \alpha$–1/T.

Let us carefully consider the temperature dependence of α for the hydrides of Pd, U, and TiMn$_{1.5}$, since the experimental data of local-mode frequencies of atoms in the crystal lattice are available for these metals, and because the hydrogen atoms occupy the only type of interstices (in Pd octahedral, in U and TiMn$_5$ tetrahedral).

Table 3.9. Constants a and b of temperature dependence of α_{AB}

Me (IMC)	Temperature range	H–T		H–D		Reference
	K	a	b	a	b	
Pd (α-phase)	195–350	0.19	−333	0.05*	−214*	[3.29]
Pd (β-phase)	175–330	−0.03	−284	−0.023	−202	[3.12]
Ti (hydride)	296–663	−0.31	−178	–	–	[3.84]
TiMn$_{1.5}$ (α-phase)	213–293	−1.435	537	–	–	[3.18]
TiMn$_{1.5}$ (β-phase)	195–296	−1.02	420	−0.75	295	[3.7]
LaNi$_5$ (β-phase)	223–323	−1.08	357	−0.53	193	[3.6]
ZrMn$_2$ (β-phase)	240–300	–	–	−0.477	240	[3.89]
ZrCr$_2$ (β-phase)	273–373	−1.90	740	−1.5	580	[3.88]

* constants a and b are found from experimental sorption isotherms for α^0_{A-B}

3.6.1 Temperature Dependence of α in the H$_2$–Pd(β-phase) System

Let us consider temperature dependences of α_{AB}, α_{BA} and α^0_{A-B}, using the experimental data and the results of quantum-statistical calculations reported in [3.12].

Most studies of the influence of temperature on isotope equilibrium deal with the H–D mixture. However, as is evident from Fig. 3.20 which shows the values of α obtained by different authors, the results differ considerably. It is demonstrated in [3.12] that neglect of the concentration dependence of the separation factor is the cause of the discrepancy. The quantum-statistical calculation of the separation factors and their temperature dependence using (3.20, 21, 23) is also reported in this work. Values of ω_H and ω_D are found from the experimental value α^{-1}_{HD} obtained at $T = 273$ K and the theorectically substantiated [3.103] condition $\omega_H/\omega_D = 1.5$, for which the calculation of protium and deuterium dissolving in the palladium β-phase agrees well with the experimental values [3.104]. The value of $\alpha^{-1}_{HT} = 2.93$ at 273 K is used for the determination of ω_T.

The frequencies calculated in [3.12] and their experimental values found by the INSS method or from the heats of solution of hydrogen isotopes in the β-phase of palladium are given in Table 3.10.

The experimental values of α for the mixtures H–D and H–T are shown in Figs. 3.21, 22. In addition, Figs. 3.20–22 present the calculation of α_{AB}, α_{BA}, and α^0_{A-B} which yields straight lines [3.12]. The computed temperature dependences of the separation factors are expressed by the equation whose constants are given in Table 3.9.

As is evident from Figs. 3.20–22 the calculated values mainly agree with the experimental ones. Appreciable deviations of the values in [3.29] are related to the fact that the influence of isotope concentration is superimposed on the temperature dependence obtained by the author. The experimental technique is chosen so that when the temperature decreases the equilibrium concentration of deuterium in the

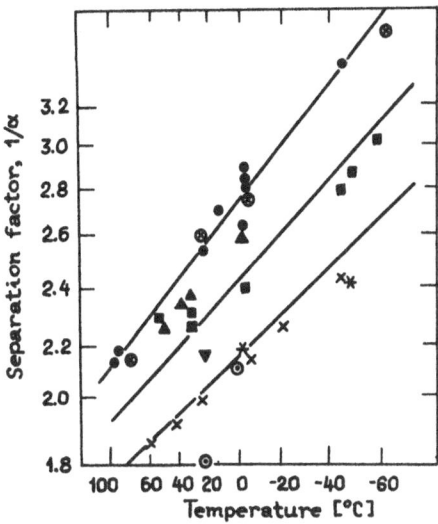

Fig. 3.20. Temperature dependence of the separation factor for the H–D mixture in H_2–Pd. $*, \blacklozenge, \bullet$: data from [3.12] at $y \approx 0, 0.5, 1$; $\blacktriangle$: data from [3.29] at $y \approx 0.6$–0.7; $\blacksquare$: data from [3.30] at $y \approx 0.5$; $\blacktriangledown$: data from [3.105] at $y < 0.5$; $\times$: data from [3.27] at $y \approx 0$; $\otimes, \odot$: chromatographic data from [3.27, 105]

Table 3.10. Local modes of isotopes of hydrogen atoms in the crystal lattice of β-phase palladium

N^o	T [K]	Local mode frequency [meV]			$\omega_H : \omega_D : \omega_T$	Reference
		ω_H	ω_D	ω_T		
1	273	59.0	39.3	31.5	1.5 : 1 : 0.8	[3.12]
2	555	54.2	36.0	–	1.5 : 1	–
3	273	56	–	–	–	–
4	70	58–59	–	–	–	–
5	298	53.3	36.3	29.3*	1.47 : 1 : 0.81	[3.17, 29]

* values obtained from the heat of tritium dissolving computed in [3.17] with the use of data on dissolving of protium and deuterium in the β-phase of palladium

gas phase increases due to the α increase, resulting in an enhancement of the observed dependence $\alpha = f(T)$.

It is interesting to note the good agreement between the experimental data and the calculation performed for the D–T mixture using two values of the separation factor (α_{HT} and α_{HD}) at definite temperature (273 K). In all cases the strongest temperature dependence of α is observed in the range of high concentration of the heavy isotope. The distinction is shown in the figures for the heats of isotope exchange reactions in the range of high and low concentrations of the heavy isotope and for the heat of the HMIE reaction. It follows from (3.16, 18, 19) that

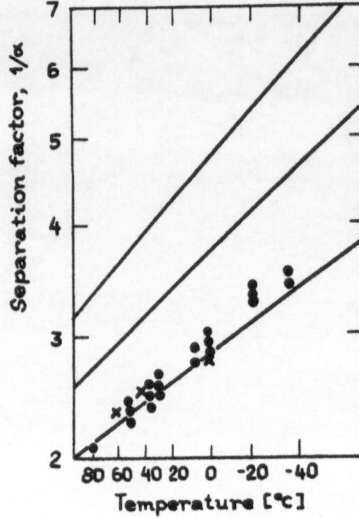

Fig. 3.21. Temperature dependence of the separation factor for the mixture H–T in H_2–Pd. •, × data from [3.32, 33]

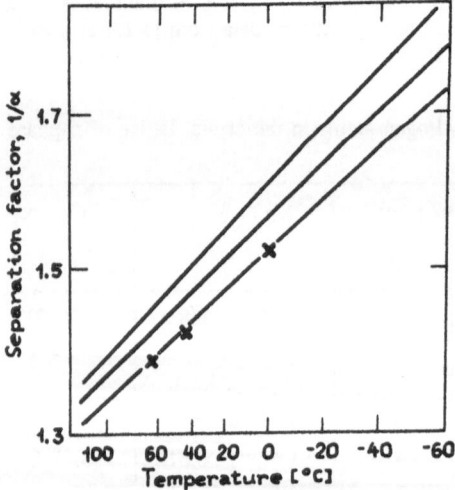

Fig. 3.22. Temperature dependence of the separation factor for the mixture D–T in H_2–Pd. × data from [3.33]

$\Delta H_{AB} = \Delta H_{BA} - \Delta H_{(AB)}$, where ΔH_{AB} is the reaction enthalpy of the HMIE reaction.

Figure 3.20 also presents the values of α obtained by the method of frontal [3.27] and displacing chromatography [3.105]. They cannot be related to a specific isotope concentration, since they are averaged values. Comparision of the α values obtained by chromatography and by the method of single adjusting for equilibrium, demonstrates the high reliability of the latter method.

Figure 3.23 [3.37] clearly illustrates the temperature dependence of the limiting separation factors α_{BA} and α_{AB} for all binary mixtures of hydrogen isotopes.

The H_2–Pd system displays an anomalous decrease of the separation factor for trace amounts of tritium [3.34], which is in no way concerned with the HMIE

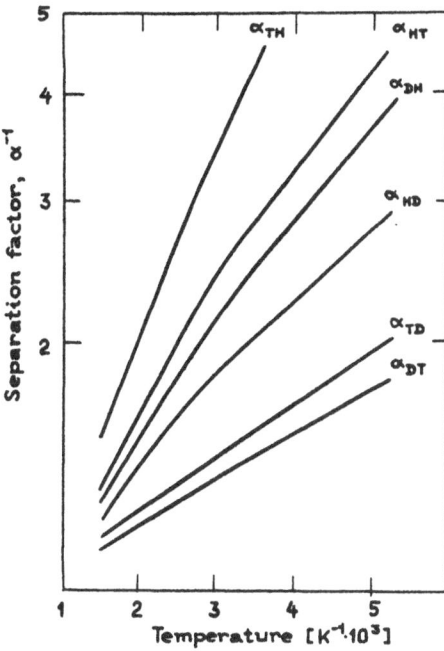

Fig. 3.23. Temperature dependencies of the separation factors for binary mixtures of hydrogen isotopes in H_2–Pd [3.37]

reaction and appears to be caused by a change in energy of hydrogen-isotope interaction in the crystal lattice of the hydride. The data on separation factors α_{HT} and α_{DT} shown in Figs. 3.21, 22 relate to an equilibrium tritium content in protium or deuterium in the gas phase of no less than 10^{-5} at%. At lower tritium concentrations, a considerable decrease of the separation factor is observed (Fig. 3.24). The figure also presents the value of α_{HT} determined for the initial tritium content in the gas of the order of 10^{-4} at% [3.32], which agrees well with the data obtained in [3.34].

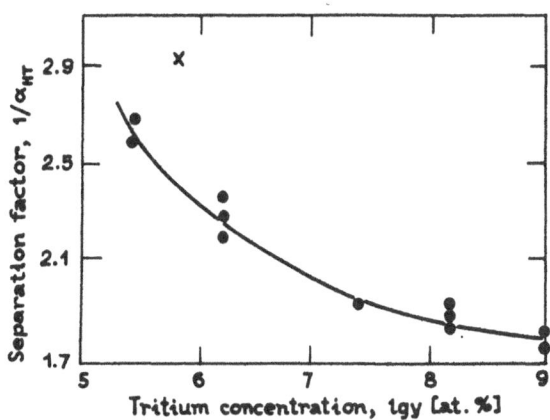

Fig. 3.24. Decrease of α_{HT} in the range of tritium trace amounts at 273 K. •, × data from [3.32, 34]

A similar α decrease is observed both for the H–T and the D–T mixture throughout the entire temperature range studied. Figures 3.25, 26 show the temperature dependences of α_{HT} and α_{DT} obtained at tritium concentrations of 7×10^{-5} and 6.5×10^{-7} at%, which are typified by substantially smaller separation factors and heats of isotope exchange reactions in comparison with the values shown in Figs. 3.21, 22. According to [3.34], the temperature dependences of α are described by the equations

$$- \ln \alpha_{HT} = 245/T - 0.23 , \tag{3.66}$$

$$- \ln \alpha_{DT} = 34.5/T - 0.16 . \tag{3.67}$$

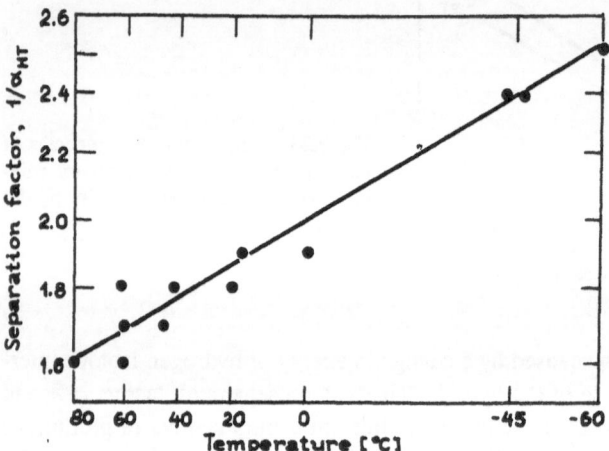

Fig. 3.25. Temperature dependence of α_{HT} in the H_2–Pd system at $y \approx 7 \times 10^{-5}$ at%

3.6.2 Temperature Dependence of α in Systems H_2–UH_3 and H_2–$TiMn_{1.5}H_{2.5}$

The experimental data for the temperature dependence of α^0_{H-D} in the H_2–UH_3 system and α_{HT} and α_{DT} for the H_2–$TiMn_{1.5}H_{2.5}$ system are presented in [3.1, 7, 10, 13, 18, 35, 106, 107].

As is shown for the H_2–UH_3 system in [3.10, 35] (see below), the separation factor α^0_{H-D} is practically independent of temperature in the temperature range 550–670 K and is equal to 0.75.

This unexpected result can be theoretically justified by means of the harmonic-oscillator model. It follows from the analysis performed in [3.21] that a slight temperature dependence is observed in this temperature interval if the frequency of local modes of hydrogen atoms in the crystal lattice falls in the range $100 < \omega_H < 125$ meV (Fig. 3.27). It is to be noted that at lower temperatures, namely, 273–373 K anomalous temperature dependence of α_{HD} is found to be possible, i.e., α increases with increasing temperature.

Investigation of the hydrogen–uranium hydride system by the neutron scattering method began in 1961 [3.108]. The first known value of $\omega_H = 92$ meV was

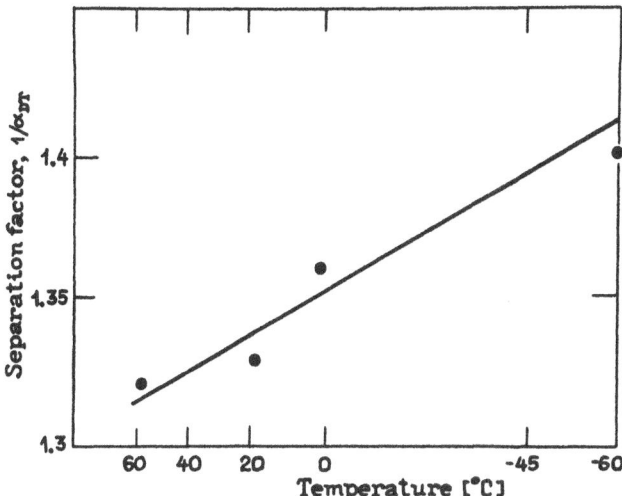

Fig. 3.26. Temperature dependence of α_{DT} in the H_2–Pd system at $y \approx 6.5 \times 10^{-7}$ at%

obtained at a temperature of 77 K; subsequent data [3.109] give $\omega_H = 112\,\text{meV}$. The latter value was obtained at 300 K, much closer to the temperature range studied. Calculation using the harmonic-oscillator model gives a similar result of $\omega_H = 113\,\text{meV}$; the value $\alpha^0_{H-D} = 0.75$ at 600 K is taken for the purpose of calculation.

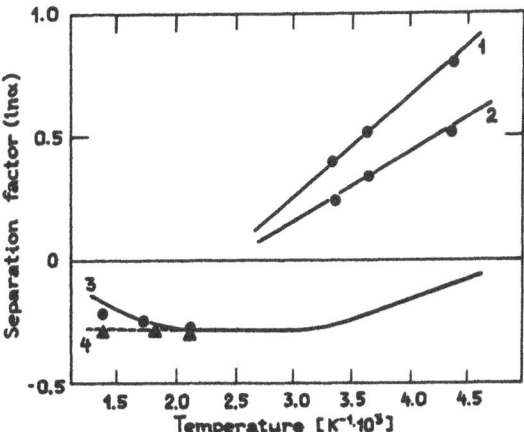

Fig. 3.27. Temperature dependence of the separation factors of hydrogen isotopes. (*1, 2*) $TiMn_{1.5}H_{2.5}$, α_{HT} and α_{HD}, respectively; (*3*) $ZrCoH_2$, α^0_{H-D}; (*4*) UH_3, α_{HD}; curve 3 is calculated at $\omega_H = 122\,\text{meV}$

As for Pd, a substantial temperature dependence is also observed for $TiMn_{1.5}$ [3.7, 13, 18]. Investigation of this system by means of INSS [3.1] gives a value for the β-phase of $TiMn_{1.5}$ of $\omega_H = 145.8$ meV, which agrees well with the calculation using the harmonic-oscillator model: $\omega_H = 145.8$ meV.

The experimental data from [3.7, 10, 13, 18] are presented in Fig. 3.27 to illustrate the different temperature dependences of α; the values of α_{H-D}^0 for uranium are recalculated for the range of low deuterium concentrations. As an example of the anomalous temperature dependence we presented the data for $ZrCoH_x$, computed from the isotherms of deuterium and protium sorption [3.87]. This dependence agrees with calculations using the harmonic-oscillator model in the temperature range 273–423 K based on the experimentally obtained value $\omega_H = 122$ meV [3.10]. This work demonstrates the possibility of hydrogen occupying interstices in the lattice of $ZrCoH_x$, resulting in a more complex temperature dependence.

3.6.3 Dependence of α on Pressure; Hydrogen Concentration in the Solid Phase

As shown in Sect. 3.5.1, when analyzing the dependence of the separation factor on pressure, it is necessary to take into account the composition of the solid phase and the positions occupied by hydrogen atoms in the lattice of the metal or IMC. In this case an effective factor $\bar{\alpha}$ must be used; see (3.57).

In real systems, the situation is complicated by the influence of chemisorbed hydrogen, which can significantly affect $\bar{\alpha}$ at low pressures. For the purpose of illustration we used the systems $Pd-H_2$ and $TiMn_{1.5}-H_2$, which are the best-studied to date. The influence of chemisorbed hydrogen on $\bar{\alpha}$ in the $Pd-H_2$ system is reported in [3.18, 28, 106, 107, 111]. Figure 3.28 presents the dependence of the separation factor α_{HT} on the fraction of chemisorbed hydrogen $1 - \varphi$.

As is evident, inversion of the isotope effect is observed at low hydrogen concentration in the solid phase; this is related to the stronger chemisorption of the heavy isotope. As is shown in [3.11], OH and OD groups are practically always present on the surface of Pd. Investigation of surface hydrogen by the INSS method [3.112] gives $\omega_{Hs} = 101$ and 120 meV. The data for the characteristic frequencies for –OH on Ni, which is similar to Pd, obtained by the same method show the presence of two values $\omega'_{O-H} = 115$ meV and $\omega''_{O-H} = 400$ meV.

Thus the energy of chemisorbed protium local modes is significantly higher than the value for local modes in the crystal lattice of Pd, which causes the positive isotope effect at $\omega_H = 2^{1/2}\omega_D = 3^{1/2}\omega_T$.

For the $H_2-TiMn_{1.5}$ system the isotope effects for adsorption and desorption are positive and there is no isotope effect inversion in the pressure dependence.

Figure 3.28 presents $\bar{\alpha}_{HT}$ as a function of $1 - \varphi$ at $T = 273$ K; the data are taken from [3.1, 7, 18, 107]. The isotherms of hydrogen sorption in the pressure range 2.6–30 mbar at temperatures of 253, 273, and 298 K are studied and the amounts of chemisorbed hydrogen are determined by the deviation from the Sieverts law in [3.18]. Further investigations corroborated the previous assumption

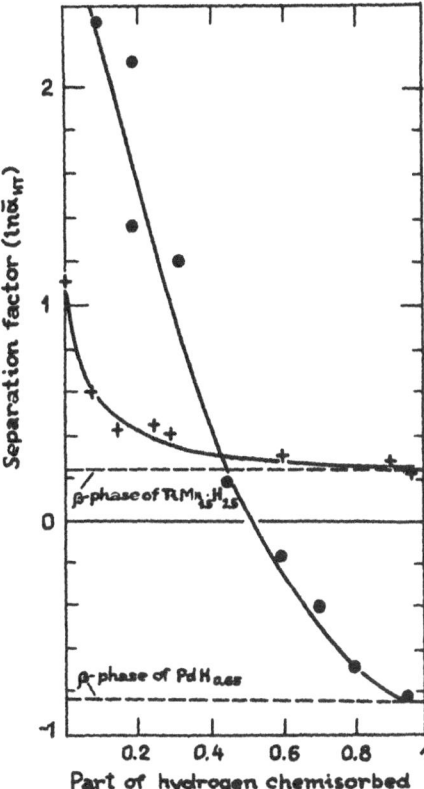

β-phase of TiMn$_{1.5}$:H$_{1.5}$

β-phase of PdH$_{0.65}$

Part of hydrogen chemisorbed

Fig. 3.28. Dependence of the separation factor $\overline{\alpha}_{HT}$ on the fraction of hydrogen chemisorbed ⊙ Pd at 353 K, + TiMn$_{1.5}$ at 273 K

about the possible segregation of Ti on the surface of IMC followed by the formation of surface hydride TiH$_x$. The presence of surface TiOH groups is shown in [3.18, 106, 114]. The study of this system by means of the INSS method began in [3.13] and was completed in [3.1, 107] showing that the calculation of $\overline{\alpha}$ using the harmonic-oscillator method in terms of different modes yields good agreement with the experimental data. Two peaks are revealed in the range of low pressure $P_{H_2} < 2$ mbar, namely, the peak $\omega_H = 160$ meV corresponding to a hydrogen atom in a tetrahedral interstice of Ti$_3$Mn and the peak $\omega_H = 74$ meV related to the modes of hydrogen atoms in the OH group.

Similar investigations for other systems are absent in the literature. It should be noted that when calculating the separation factors in the transition range α–β and in the range of high pressures using the harmonic-oscillator model, it is necessary to know not only the local mode frequencies but also the fraction of hydrogen atoms in either tetrahedral or octahedral interstices. Many studies have been devoted to the structure of IMC hydrides and the determination of hydrogen atom positions in them. However, quantitative estimation of $\overline{\alpha}$ is often complicated by the absence of data on α and on frequencies of local modes in the same temperature ranges. From the practical standpoint, determination of $\overline{\alpha}$ by the method of single adjusting for equilibrium is more straightforward.

3.7 Interrelation Between Isotope Effects in H–T, H–D, and D–T Mixtures and Distribution of Tritium in Three-Isotope Mixtures H–D–T Between Equilibrium Phases

Since hydrogen has three isotopes, the separation factor for a mixture that has not been studied can be computed from the values of α known for two isotope pairs. However, the simplest and most obvious relation

$$\alpha_{HT} = \alpha_{HD}\alpha_{DT} \tag{3.68}$$

appears to be unsuitable in systems of substances in which the molecules on only one of them involve two or more exchanging atoms, because of nonequiprobable isotope distribution in the HMIE reactions. In its general form this relation with regard to the HMIE reactions in both exchanging substances is considered in [3.15, 26]. In systems with hydride phases of metals and IMC this relation, along with the concentration dependence of α, is significantly simplified.

Let us consider the reactions of complete isotope substitution for all the isotope mixtures of hydrogen:

$$2H(Me) + T_2 = 2T(Me) + H_2 , \tag{3.69}$$

$$2H(Me) + D_2 = 2D(Me) + H_2 , \tag{3.70}$$

$$2D(Me) + T_2 = 2T(Me) + D_2 . \tag{3.71}$$

Since the reaction (3.69) can be presented as the sum of reactions (3.70) and (3.71), $\alpha^0_{H-T} = \alpha^0_{H-D}\alpha^0_{D-T}$ or, with regard to (3.46),

$$\alpha_{HT} = \alpha_{HD}\alpha_{DT}\left(\frac{K_{HD}K_{DT}}{4K_{HT}}\right)^{1/2} , \tag{3.72}$$

The error caused by neglect of the last factor in (3.72) increases with decreasing temperature (e. g., when the temperature falls from 400 to 200 K, the error more than doubles and is equal to $\approx 17\%$).

Expression (3.72) is a rigorous thermodynamic relationship, which can be used both for the calculation of α in the unstudied, for instance D–T, mixture (for this reason, generally, only the dependences for α_{HT} and α_{HD} are considered here) and for checking and correcting the obtained values of α_{AB}. So, for instance, when computing by (3.72), the values $\alpha^{-1}_{HD} = 2.16$, $\alpha^{-1}_{DT} = 1.47$ used in the calculation of the concentration dependences in Fig. 3.13 at $T = 273$ K and at $K_{HT} = 2.43$, $K_{HD} = 3.19$, $K_{DT} = 3.79$ lead to $\alpha^{-1}_{HT} = 2.85$, which is practically equal to the value obtained from (3.63) for the α_{HT} temperature dependence.

Let us give one more example. As noted above, the value $\alpha^0_{H-D} = 1.5$ at 313 K is computed from the isotherms of protium and deuterium sorption on vanadium (V). Since $\alpha_{HD} = \alpha^0_{H-D}(K_{HD}/4)^{1/2}$, one finds $\alpha_{HD} = 1.36$. On the other hand, it is found from (3.72) that at $K_{HT} = 2.66$ and $K_{DT} = 3.84$, $\alpha_{HD} = 1.46$, i. e., it differs slightly from the value computed from the isotherms of protium and deuterium sorption.

Due to an additional separation effect resulting from the HMIE reactions of hydrogen, especially in systems where an isotope effect inversion can occur upon a change of isotope concentration, an extraordinary relation exists between the separation factors for different mixtures of hydrogen isotopes, which, at first sight contradicts the popular opinion in the theory of isotope effects.

Let us make this consideration clear using the example of the above-considered concentration dependence of $\alpha_{A–B}$ for the H_2–LaNi$_5$ system shown in Fig. 3.14. As noted above, for the mixtures H–T and H–D a strong concentration dependence of α is observed. Since $\alpha_{H–T}$ depends more strongly on concentration than $\alpha_{H–D}$, at 195 K and for a protium concentration of $\approx 30\%$ these separation factors are found to be equal and at lower protium concentration the separation factor of the H–T mixture (these isotopes differ the most in mass) appears to be lower than for the H–D mixture.

At 273 K the pattern is found to be more complicated because of the isotope effect inversion. A considerable isotope effect for the H–T mixture corresponding to greater difference in mass is observed over a rather narrow concentration range: in the range of high protium concentrations (> 85 at%) and at protium concentrations lower than ≈ 40 at%. In the range from 85 to 60 at% $\alpha_{HD} > \alpha_{HT}$ and finally, from 60 to 40 at% the isotope effect is positive in the H–D mixture and negative in the H–T mixture. For the D–T mixture throughout the concentration range the isotope effect is negative. Hence, the negative isotope effect in the D–T mixture is higher than in the H–T mixture in the concentration range from isotope effect inversion in the H–T mixture to the intersection of the curves $\alpha_{H–T}$ and $\alpha_{D–T}$.

From the above considerations it follows that the terms accepted in literature, namely, positive and negative (normal and anomalous) isotope effects are unique only in systems with considerable isotope effects, whereas, systems with concentration inversion of the isotope effect hold an intermediate position.

In addition to the separation of binary mixtures of tritium with protium or deuterium, one must consider the task of tritium extraction from ternary mixtures H–D–T, in which the ratio between protium and tritium can vary over a wide range. In the general case ternary mixture separation can be solved with the use of the theory of multicomponent mixtures separation. In many cases the task of isotope extraction from a multicomponent mixture can be reduced to the separation of two isotopes at a constant content of the third component (e. g., at low tritium content, its concentration or depletion at constant ratio of the other isotopes). In this case the data on equilibrium tritium distribution between gas and solid phases are sufficient.

When all hydrogen isotopes are present, the tritium distribution between equilibrium phases is expressed by the equation

$$\alpha_T = \frac{x_T(1 - y_T)}{y_T(1 - x_T)} \frac{[T(Me)]}{[H(Me) + D(Me)]} \frac{2[H_2] + 2[HD] + 2[D_2] + [HT] + [DT]}{[HT] + [DT] + 2[T_2]}. \tag{3.73}$$

The work [3.115] is devoted to the concentration dependence of α_T. The introduction of equilibrium constants of the HMIE reactions and substitution of isotope concentrations in the metal by the separation factors for binary mixtures enables one to bring (3.73) into the form

$$\alpha_T = \frac{1 + x + 0.5y + 0.5xy(K_{DT}/K_{HT}K_{HD})^{1/2} + x^2/K_{HD}}{0.5 + 0.5x(K_{DT}/K_{HT}K_{HD})^{1/2} + y/K_{HT}} \frac{\alpha_{HD}^{-1}}{\alpha_{HT}^{-1}(2\alpha_{HD}^{-1} + x)}, \tag{3.74}$$

where $x = [HD]/[H_2]$, $y = [HT]/[H_2]$.

The dependence of α_T on the concentration of protium and deuterium in the gas phase, as well as the concentration dependence of the separation factor α_{A-B}, differs from the additive one due to the deviation of the HMIE reaction equilibrium constants from $K^\infty = 4$.

It follows from (3.74) that α_T depends on temperature (in terms of equilibrium constants of the three HMIE reactions and separation factors for binary mixtures) and on the isotope composition of hydrogen characterized by the values of x and y. The latter values can be expressed in terms of the protium concentration in the binary isotope mixtures H–D (C^{HD}) and H–T (C^{HT}) using relations analogous to (3.44):

$$\left(1 - C^{HD}\right) x^{-2} + \left(0.5 - C^{HD}\right) x^{-1} - C^{HD}/K_{HD} = 0, \tag{3.75}$$

$$\left(1 - C^{HT}\right) y^{-2} + \left(0.5 - C^{HT}\right) y^{-1} - C^{HT}/K_{HT} = 0. \tag{3.76}$$

Values of α_T can be found for any composition of the ternary mixture by solving the set of equations (3.74–76).

In the range of low tritium content, assuming $[H_2] + [HD] + [D_2] \approx 1$ and $[H(Me)] + [D(Me)] \approx 1$, instead of (3.73) we can write

$$\alpha_T = \frac{[T(Me)]}{0.5[HT] + 0.5[DT]}; \tag{3.77}$$

and instead of (3.74) we obtain

$$\alpha_T = \frac{\alpha_{HT}[H(Me)]}{([H_2] + [HD])(K_{DT}/K_{HT}K_{HD})^{1/2}}. \tag{3.78}$$

The values $[H_2]$ and $[HD]$ can be found from (3.75), $[H(Me)]$ from α_{H-D} computed by (3.43).

As an example, the dependencies of α_T^{-1} on the concentration of protium (C) in the gas phase calculated using (3.78) for three temperatures are presented in Fig. 3.29 for the H_2–Pd system. It is evident from Fig. 3.29 that the largest deviations from the additive dependence described by equation

$$\alpha_T^{-1} = \alpha_{HT}^{-1}C + \alpha_{DT}^{-1}(1 - C) \tag{3.79}$$

are observed at 233 K; the deviations fall with increasing temperature.

Expressions for the binary separation factors in the ternary mixture of hydrogen isotopes in terms of the isotope composition of hydrogen in the solid phase, equilibrium constants of the HMIE reactions, and the ratios of equilibrium pressures of hydrogen isotopes over the solid phase are obtained in [3.116]. Since $P_{A_2}/P_{B_2} = (\alpha_{A-B}^0)^2$, the equations obtained in this work can be represented as

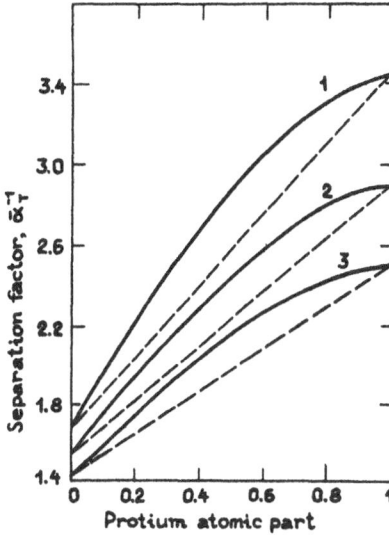

Fig. 3.29. Dependence of the separation factor α_T in system H_2–Pd in the ternary mixture H–D–T on the composition of the gas phase. (*1*) 233 K, (*2*) 273 K, (*3*) 313 K

$$\alpha_{H-T} = \left(\alpha_{H-T}^0\right)^2 \frac{x_H + \left(x_D/\alpha_{H-D}^0\right)\left(K_{HD}/4\right)^{1/2} + \left(x_T/\alpha_{H-T}^0\right)\left(K_{HT}/4\right)^{1/2}}{x_H\alpha_{H-T}^0\left(K_{HT}/4\right)^{1/2} + x_D\alpha_{D-T}^0\left(K_{DT}/4\right)^{1/2} + x_T},$$

(3.80)

$$\alpha_{H-D} = \left(\alpha_{H-D}^0\right)^2 \frac{x_H + \left(x_D/\alpha_{H-D}^0\right)\left(K_{HD}/4\right)^{1/2} + \left(x_T/\alpha_{H-T}^0\right)\left(K_{HT}/4\right)^{1/2}}{x_H\alpha_{H-T}^0\left(K_{HT}/4\right)^{1/2} + \left(x_T/\alpha_{D-T}^0\right)\left(K_{DT}/4\right)^{1/2} + x_D},$$

(3.81)

$$\alpha_{D-T} = \left(\alpha_{D-T}^0\right)^2 \frac{x_D + \left(x_H\alpha_{H-D}^0\right)\left(K_{HD}/4\right)^{1/2} + \left(x_T/\alpha_{D-T}^0\right)\left(K_{DT}/4\right)^{1/2}}{x_H\alpha_{H-T}^0\left(K_{HT}/4\right)^{1/2} + x_D\alpha_{D-T}^0\left(K_{DT}/4\right)^{1/2} + x_T}.$$

(3.82)

where x_H, x_D, and x_T are the atomic fractions of protium, deuterium, and tritium in the solid phase.

4. Kinetics of Hydrogen Isotope Interaction with Hydride Phases

4.1 Equation of Formal Kinetics

Isotope exchange reactions differ from ordinary chemical reactions in that the concentrations of the reagents (in the case in question hydrogen pressure and its content in the hydride phase) remain unchanged and isotope distribution between the components is the sole result of the process. In this case the process of isotope exchange occurs at any isotope distribution with constant rate. As long as the isotope distribution does not attain equilibrium, determined by the separation factor α, this exchange results in a change of isotope concentrations x and y in the hydride and in the gas phase. When isotope equilibrium is established, exchange continues with the same rate but invisibly (i.e., it no longer results in a change of isotope concentrations).

The rate of reaction depends, apart from temperature, on the concentration of the exchanging substances and, for the case of hydrogen exchange with hydride phases of metals and IMC, can be represented as

$$R = K P^m n^q \tag{4.1}$$

where K is the rate constant.

Since the thermodynamic isotope effect is determined by the ratio of direct and reverse reactions of isotope exchange, $\alpha = \overrightarrow{R}/\overleftarrow{R} = \overrightarrow{K}/\overleftarrow{K}$ the enthalpy of any reversible reaction is equal to the difference of activation energies of direct and reverse reactions, and, at $\alpha \neq 1$,

$$\overleftarrow{E}_{act} = \overrightarrow{E}_{act} + \Delta H \tag{4.2}$$

where ΔH is the enthalpy of the isotope exchange reaction. The difference in the activation energies can be essential in the hydrogen isotope exchange reaction, which is characterized by a considerable temperature dependence of α.

Formally, the overall order and molecularity of the reactions of hydrogen isotope exchange with hydride phases, as for any isotope exchange reaction, is equal to 2. However, as a rule, the reactions proceed through a number of intermediate steps and hence, the order (equal to $m + q$) can deviate from this value and may also be fractional.

A simple exponential equation is usually used to describe the formal kinetics of isotope exchange reactions. It includes a single constant (rate of exchange R)

and is valid for the case of uncomplicated isotope exchange (e. g., in the absence of diffusion deceleration during the transport of exchanging substances in the reaction zone) occurring if only one of the following conditions holds

1. a minor isotope effect in the system ($\alpha \approx 1$);
2. a low concentration of one of the isotopes ($x, y \ll 1$);
3. a small amount of one of the reagents (hydrogen in gas or solid phase).

Consider the isotope-exchange reaction between hydrogen gas and hydride phase of metal or IMC in the general form.

$$A(Me) + AB \leftrightarrow B(Me) + AA. \tag{4.3}$$

Since the rate of isotope exchange is determined by the concentration change in the gas phase, the kinetic equation can be written in differential form in terms of the concentration y

$$-N_g \frac{dy}{d\tau} = \overrightarrow{R} S(1 - x)y - \overleftarrow{R} S x(1 - y), \tag{4.4}$$

where N_g, N_s are the number of hydrogen moles in gas and solid phases; S is the surface of contact of phases (surface area of the solid phase); $\overrightarrow{R}$ and $\overleftarrow{R}$ are the rates of direct and reverse reaction in mol/(m^{2}s). For this purpose the concentration x can be obtained from the mass balance for the heavy isotope

$$x = x_\infty - N_g/N_s(y - y_\infty). \tag{4.5}$$

When the condition $\alpha \approx 1$ holds, considering $\overrightarrow{R} = \overleftarrow{R}$, we obtain

$$-N_g \frac{dy}{d\tau} = RS(y - x), \tag{4.6}$$

Substituting the concentration x from (4.5) into (4.6) we obtain

$$\frac{-dy}{y - y_\infty} = RS \left(\frac{1}{N_g} + \frac{1}{N_s} \right) d\tau. \tag{4.7}$$

Integration leads to the simplest exponential law of exchange

$$-\ln(1 - F) = RS \left(\frac{1}{N_g} + \frac{1}{N_s} \right) d\tau = r\tau \tag{4.8}$$

or

$$F = 1 - \exp(-r\tau) \tag{4.9}$$

where F is the exchange degree given by $F = (y_0 - y)/(y_0 - y_\infty)$, r is the observed rate constant (s^{-1}).

According to (4.8) the kinetics of the isotope exchange reaction as plot $-\ln(1 - F)$ vs. τ can be represented as a straight line passing through the origin and having a slope equal to r (Fig. 4.1). The slope, of the line depends not only on

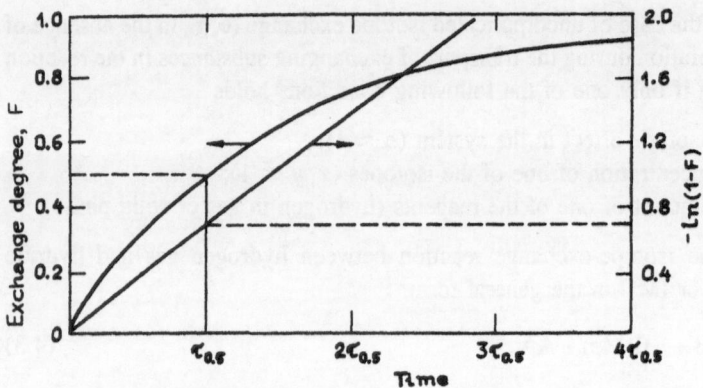

Fig. 4.1. Kinetics of uncomplicated isotope exchange

the rate of isotope exchange but also on the hydrogen amount in the gas and solid phases of the system. The half-exchange time $\tau_{0.5}$ is often taken as a characteristic of the exchange kinetics:

$$\tau_{0.5} = \ln 2/r = 0.693 \Big/ \left[RS \left(\frac{1}{N_g} + \frac{1}{N_s} \right) \right]. \tag{4.10}$$

Depending on the condition of exchange (number of moles N_g and N_s) even at a constant rate of exchange, the half-exchange time can vary over a wide range. That is why one should consider the relative character of this characteristic of isotope exchange kinetics.

For the condition $x, y \ll 1$, equation (4.4) is written in the form

$$-N_g \frac{dy}{d\tau} = \vec{R} S \left(y - \frac{x}{\alpha} \right). \tag{4.11}$$

Similar transformations lead to the equation

$$-\ln(1 - F) = \vec{R} S \left(\frac{1}{N_g} + \frac{1}{\alpha N_s} \right) \tau = r\tau. \tag{4.12}$$

In the range of low content of the heavy isotope B, regardless of the direction of the reaction, and in coordinates $-\ln(1 - F)$ vs. τ, the kinetics of exchange is described by a straight line, the slope of which depends not only on N_g and N_s but also on the separation factor α.

In the range of high contents of the heavy isotope when $\alpha = (1 - y)/(1 - x)$, the kinetic equation transforms to

$$-\ln(1 - F) = \vec{R}^* S \left(\frac{1}{N_s} + \frac{1}{\alpha N_g} \right) \tau = r^*\tau. \tag{4.13}$$

At equal rate of exchange $(R = R^*)$ the experimental exchange rate constants in the range of low and high contents of the heavy isotope only coincide if $N_g = N_s$.

The differences between r and r^* increase with increasing deviation of the ratio N_s/N_g from 1, attaining a maximum value when the change of isotope composition in one of the phases can be ignored (for $N_g \ll N_s$ $r/r^* = \alpha$, and for $N_g \gg N_s$ $r/r^* = \alpha^{-1}$). Thus the half-exchange time depends not only on the amount of hydrogen in the exchanging phases and on the separation factor, but also on the range of isotope concentration change as well.

Finally, in the case of the third condition, which is of practical interest only at $N_s \gg N_g$ ($x = x_0 = x_\infty = $ const), equation (4.4) transforms to the relation

$$\frac{-N_g}{\alpha - \varepsilon x_0} \frac{dy}{d\tau} = \overrightarrow{R} S \left(\frac{y - x_0}{\alpha - \varepsilon x_0} \right). \tag{4.14}$$

Integration of (4.14) with the substitution $y_0 = x_0/(\alpha - \varepsilon x_0)$ results in the final expression:

$$- \ln(1 - F) = \frac{\overrightarrow{R} S \tau (\alpha - \varepsilon x_0)}{\alpha N_g} = r\tau. \tag{4.15}$$

Thus, for small amounts of one of the exchanging substances (i. e., hydrogen) not only the separation factor influences the dependence of exchange degree on time but the amount of gaseous hydrogen in the system and the initial content of isotopes in the hydride phase have an effect too.

Often in the above equations of exchange kinetics the surface is expressed in terms of a specific surface area (S_s), which is related to the mass g of the solid phase by $S_s = S/g$. In the latter case, (4.8, 12, 15) are written as:

at $\alpha = 1$,

$$- \ln(1 - F) = Rg \left(\frac{1}{N_g} + \frac{1}{N_s} \right) \tau; \tag{4.16}$$

at $\alpha \neq 1$ and $x, y \ll 1$

$$- \ln(1 - F) = \overrightarrow{R} g \left(\frac{1}{N_g} + \frac{1}{\alpha N_s} \right) \tau; \tag{4.17}$$

at $\alpha \neq 1$ and $N_g \ll N_s$

$$- \ln(1 - F) = \frac{\overrightarrow{R} g (\alpha - \varepsilon x_0)}{\alpha N_g} \tau. \tag{4.18}$$

Recall that in the above-listed relations the value $\overrightarrow{R}$ is the rate of reaction (4.3) in the direction from left to right, which results in a positive isotope effect. If the heavy isotope is concentrated in the gas phase (negative isotope effect), reaction (4.3) must be written as follows: $A_2 + B(Me) \leftrightarrow AB + A(Me)$ and $AB + B(Me) \leftrightarrow B_2 + A(Me)$ for the range of low and high contents of the heavy isotope B. The value $\overrightarrow{R}$ in (4.12, 13, 15, 17, 18) is to be understood as the rate of these reactions in the opposite direction.

The relations considered are characteristic for the kinetics of uncomplicated isotope exchange, when all hydrogen atoms in the hydride phase (as well as in a hydrogen molecule) are equal and the rate of reagents arriving in (and leaving) the reaction zone is sufficiently high so as not to effect the isotope exchange kinetics. However, in many systems with hydride phases of metals and IMC, the kinetics of heterogeneous isotope exchange is totally determined by diffusion processes, whose behaviour differs from chemical kinetics, because of the low diffusion coefficients of hydrogen atoms in the solid phase. The most difficult task arises when differential equations of chemical kinetics and diffusion transport of substance are to be solved simultaneously.

4.2 Mass-Transfer of Hydrogen in Systems with Hydride Phases

Mass-transfer in systems with a solid phase proceeds under the condition of constancy of the contact surface, which does not depend on the hydrodynamic operating mode of the apparatus. However, in spite of this fact, the behaviour of mass-transfer in the general case remains complex due to the large number of steps in the mass-transfer process in the solid phase and the influence of the amount of dissolving hydrogen on the rate of the step corresponding to diffusion in the solid phase.

It is evident that for hydrogen isotope exchange the influence of isotope composition on the parameters of the mass-transfer process can be neglected, i. e., the efficiency of the process is assumed to be constant along the whole height of the separation column. Let us consider the main features of mass-transfer for hydrogen isotope exchange proceeding under stationary conditions in a separation column in terms of this simplification.

Since the rigorous solution of the total problem of mass-transfer has not been obtained to date, we consider the agreed-upon approximation of additive contributions of axial dispersion and of mass-exchange processes in the gas and solid phases to the mass-transfer resistance [4.1–3]:

$$\frac{1}{K_{0g}} = \frac{1}{\beta_g} + \frac{1}{m\beta_s} + \frac{D_{eff}}{w^2}, \qquad (4.19)$$

where β_g and β_s are the coefficients of mass-exchange in the gas and solid phases, respectively; K_{0g} is the coefficient of mass-transfer for the gas phase; D_{eff} is the effective coefficient of axial diffusion; w is the linear gas velocity; m is a coefficient equal to the slope of the tangent to the equilibrium line ($m = dx/dy$, where x, y are isotope concentrations).

When considering mass-transfer in the range of low concentrations of the heavy isotope, the coefficient m is equal to the separation factor α.

From (4.19), the height of the transfer unit (HTU) for the gas phase depends not only on the rate of interphase exchange but also on the intensity of the axial dispersion

$$h_{0g} = h_g + \frac{\lambda h_s}{\alpha} + h_{ad} , \qquad (4.20)$$

where $h_g = G_{sp}/\beta_g a_{gr}$, $h_s = L_{sp}/\beta_s a_{gr}$, $h_{ad} = D_{eff}/w = D_{eff}S_0\varrho/mG_{sp}$, G_{sp} and L_{sp} are specific molar flows (molar flow densities) of gaseous hydrogen and hydrogen in the solid phase; a_{gr} is the contact surface (the surface area of grains) per unit volume of column; S_0 is the fraction of free section (space) of the column; ϱ is the gas density; m is the hydrogen molecular mass, and λ is the flow ratio $(\lambda = G_{sp}/L_{sp})$.

The component of HTU corresponding to the axial dispersion depends on the structure of flows in the apparatus and is responsible for the scaling effect, which effects an increase in HTU when laboratory columns are replaced by industrial ones. The effective coefficient of axial diffusion D_{eff} is used as a quantitative characteristic of the axial dispersion. It depends on the effects of both axial dispersion determined by molecular, turbulent, and convective mixing processes and lateral irregularities, whose influence lessens as the lateral dispersion increases.

In columns with an immobile bed of solid phase the last effect is connected with the lateral distribution (profile) of gas flow rates. The introduction of D_{eff} including all hydrodynamic effects enables one to describe in a one-dimensional approximation the lateral irregularities as a rise of the axial dispersion [4.3]. It is evidently difficult to calculate D_{eff} since it depends not only on the geometry and operating mode of the apparatus (loading, temperature, pressure and so on) but also on the sizes and shape of the particles, the character of their polydispersity and even the method of column filling. That is why the experimental determination of D_{eff} is currently the most reliable.

The dependences of D_{eff} and h_{ad} on the linear gas velocity in a column of diameter 1.5 cm filled with a solid phase with grains of size 2–3 mm, which are presented in Fig. 4.2, reflect the effect of axial dispersion in hydrogen on HTU in a column with an immobile bed of solid phase.

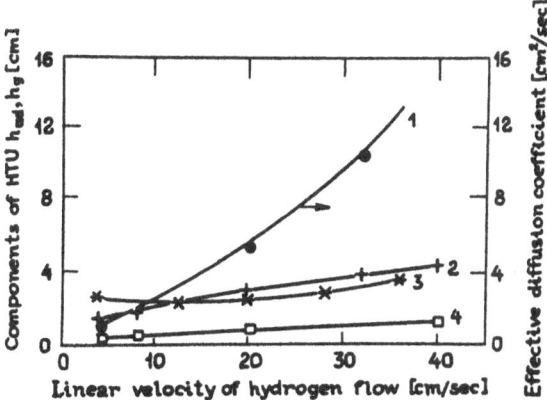

Fig. 4.2. Influence of linear hydrogen velocity on the effective coefficient of axial diffusion (*1*) and on components of HTU h_{ad} at 77 K (*3*), h_g at 77 K (*2*), and at 293 K (*4*) in a column filled with spherical grains of sorbent of diameter 2.5 mm

The values of D_{eff} are calculated from the experimental response curves (c-curves) obtained at atmospheric pressure [4.4]. The experiments carried out at 77 and 298 K show that D_{eff} does not depend on temperature at gas velocities over 0.04 m/s. Only at lower gas velocity, when D_{eff} slightly exceeds the coefficient of molecular diffusion, the temperature does effect the axial dispersion. So, at $w > 0.04$ m/s, the contribution of the axial dispersion to HTU does not depend on temperature and, in the column filled with spherical grains of size 2–3 mm, is equal to 2–4 mm at $0.01 < w < 0.4$ m/s. As is shown in [4.4, 5], D_{eff} decreases with reduction of the grain size and increases when the column is filled with grains of irregular shape.

The external diffusion resistance depends strongly on the hydrodynamic mode of gas flow. At the transition from a laminar flow pattern to a turbulent one β_g increases and the contribution of external diffusion resistance falls. The following equation is usually used for describing the experimental data on mass-exchange in the range of external diffusion

$$\text{Nu}_g = A \, \text{Re}_g{}^m \, \text{Pr}_g{}^n, \tag{4.21}$$

where A, m, and n are constants.

For hydrogen over a wide range of temperatures and pressures the physical properties change so that the Prandtl criterion is practically equal to 1. Hence, the equation of mass-exchange in gas is written

$$\text{Nu}_g = A \, \text{Re}_g{}^m, \tag{4.22}$$

where $\text{Re} = w d_e \varrho / \mu$; $\text{Nu} = \beta_g^c d_e / D$; D is the coefficient of molecular diffusion; d_e is the equivalent size of channels in a granular bed; μ is the gas viscosity; and β_g^c is the coefficient of mass-exchange in the gas phase ($\beta_g = \beta_g^c \varrho / RT$).

Since the geometrical surface of spherical grains of equal size in the unit volume is equal to:

$$a_{\text{gr}} = \frac{6(1 - S_0)}{d_{\text{gr}}} \quad \text{and} \quad d_e = \frac{4 S_0}{a_{\text{gr}}}, \tag{4.23}$$

from (4.22) and since $h_g = G_{\text{sp}} / \beta_g a_{\text{gr}} = w S_0 / \beta_g^c a_{\text{gr}}$ we obtain

$$h_g = \frac{1}{6 A D} \left(\frac{2}{3} \right)^{1-m} \left(\frac{S_0 d_{\text{gr}}}{1 - S_0} \right)^{2-m} \left(\frac{\mu}{\varrho} \right)^m w^{1-m}. \tag{4.24}$$

For a laminar flow the index m is close to 0.5 and for a turbulent flow it can reach 0.8–0.9 [4.6, 7].

The dependences of the component h_g corresponding to the external diffusion on hydrogen velocity at atmospheric pressure and temperatures of 77 and 273 K are computed from (4.24) at $A = 0.725$ and $m = 0.7$ for spherical grains of diameter 2.5 mm ($S_0 = 0.48$). As is evident from the dependences presented in Fig. 4.2, this component of HTU at 273 K does not exceed 1 mm, i. e., it is several times smaller than the linear size of sorbent grains and at 77 K it increases by a factor of 3–4 (primarily due to the reduction of the diffusion coefficient by a factor of 10); the coefficient of mutual diffusion for H_2–D_2 at 273 K is equal to 1.13 cm²/s and at 77 K it is 0.116 cm²/s. It reaches a value h_g = 4 mm at w = 0.4 m/s. It should be noted that considering the effect of temperature on h_g at G_{sp} = const, the distinctions appear to be less considerable (since at w = const an increase of G_{sp} by a factor of 3.5 corresponds to a decrease in temperature from 273 to 77 K). A rather weak dependence of h_g on the loading in the column, which becomes less appreciable at the transition to the turbulent flow pattern, is to be noted.

It follows from the analysis of (4.24) that, at G_{sp} = const, h_g does not depend on pressure, since $D \propto p^{-1}$, $\varrho \propto p$, and the hydrogen viscosity is practically independent of pressure (e. g., when the hydrogen pressure changes from 0.1 to 10 MPa, it increases by only 2 %).

A reduction in size of the sorbent grains is assumed to result in a decrease of h_g, which can be more considerable for the laminar flow pattern for hydrogen (provided the change of free space in the column can be ignored when filling with sorbent grains of different size, $h_g \propto d_{gr}^{2-m}$).

A feature of the internal mass-exchange coefficient β_s is its independence of the hydrodynamic pattern in the column. Hence, the corresponding component h_s increases linearly with the degree of loading in the column. In h_s–G_{sp} coordinates the dependence is described by a linear law, whose slope is determined by the relation $\lambda/(\alpha\beta_s a_{gr})$, i. e., the dependence of HTU on loading becomes weaker when the rate of this transfer step or the contact surface increases. If the grain size does not effect β_s, in terms of (4.23) a reduction of the grain size leads to a proportional decrease of h_s.

For the components h_{ad} and h_g we ignored the effect of hydrogen isotope composition on their physical properties, i. e., we assumed that these components are equal for exchange between any hydrogen isotopes. When considerable isotope effects exist in a system (just such systems are of interest here), it should be taken into account that if the heavy isotope is concentrated in the solid phase, the contribution of h_s to HTU for the gas decreases rapidly with increasing separation factor. For example, in systems with IMC the h_s contribution is smallest for H–T isotope exchange (since $\alpha_{HT} > \alpha_{HD} > \alpha_{DT}$).

Solving Fick's second law and assuming spherical grains, and a linear character of the equilibrium equation, one obtained the relation [4.8]

$$\beta_s^c a_{gr} = \frac{4\pi^2 D_e}{d_{gr}^2} \qquad (4.25)$$

where D_e is the effective diffusion coefficient in the sorbent.

In spite of a number of rough assumptions this expression is corroborated by practice.

Taking the relation $\beta_s = \beta_s^c E_H$ (E_H is the hydrogen content in the sorbent in mol/m³) into account we obtain:

$$h_s = \frac{G_{sp} d_{gr}^2}{4\pi^2 D_e E_H} . \qquad (4.26)$$

The character of the temperature and pressure influence on h_s depends on the processes going on in the sorbent grain, i. e., specific properties of the solid phase including the nature of the isotope effect, the mechanism and steps of the isotope exchange, affect the efficiency of mass-transfer only through the component of HTU corresponding to internal diffusion.

4.3 Experimental Methods for Studying the Kinetics of Hydrogen Isotope Interaction with Hydride Phases

To study the kinetics of isotope exchange reactions of hydrogen with hydrides of metals and IMC one may use the same methods as used for investigating ordinary chemical reactions. In isotope exchange the following parameters remain constant: the total number of molecules, the chemical composition of substances, and molecular interaction; hence, one may assume that $\Delta V = 0$. This means that the reaction occurs not only at T = const (since the heat effect is slight) but under isobaric conditions as well, a fact that simplifies the experimental data processing.

There are three types of methods for studing isotope exchange reactions: static, flowing, and circulatory-flowing. The static method with forced gas circulation has the widest application in studies of the kinetics of hydrogen isotope interaction with hydride phases.

The experimental apparatus is shown schematically in Sect. 3.2. The experimental technique is similar to the above-mentioned one used for α determination and, as a rule, the experiments to determine of α and R were carried out simultaneously.

In the case of low heavy isotope concentrations the kinetics is expressed either by (4.8) at $\alpha \approx 1$ as is observed for a number of IMC (e. g., TiFe, TiMo) or by (4.12) when α differs markedly from 1. It is to be noted that, at temperatures higher than 700 K, the isotope exchange kinetics is described by (4.8) for practically all hydrogen-metal–(IMC)-hydride systems.

The possibility of studying isotope exchange reaction kinetics not only on granulated but also on powdered sorbents (e. g., samples of activated IMC) is the advantage of the static method. When the gas diffusion in the sorbent grains leads to a deviation from first-order kinetics, the rate constant is determined by differential forms of (4.7.11).

Contrary to methods of studying isotope exchange reaction kinetics where the rate constant, which is as well a fundamental characteristic of the isotope exchange reaction as the separation factor, is determined, results for mass-transfer efficiency in a column depend on the details of its construction and on the hydrodynamics of flows. In this case the column can operate both under conditions of solid phase movement and with an immobile bed. The noted reasons not only affect HTU components (Sect. 4.2) but can result in different additional effects of the isotope mixing, which are characteristic for the concrete apparatus and influence the efficiency of its operation.

The simplest and most commonly used method (only for gas-liquid systems) is based on the realization of a steady operating regime without withdrawal of product ($\lambda = 1$) with the subsequent measurement of the separation degree K attained and calculation of HTU by the Fenske-Underwood equation

$$K = \exp[(\alpha - 1)N_{0s}] = \exp\left[\frac{(\alpha - 1)N_{0g}}{\alpha}\right] \tag{4.27}$$

where $N_{0s} = H/h_{0s}$ and $N_{0g} = H/h_{0g}$; H is the column height.

Practical realization of this method in systems with solid phase is complicated by the difficulties of realizing continuous counter-current separation (chap. 5).

Hence, for the study of interphase mass-transfer efficiency in systems with solid phase, one uses techniques based on the passage of gas through the column with immobile sorbent bed and on the measurement of isotope distribution in time in the output hydrogen flow.

The majority of the experimental data, presented below on the mass-transfer efficiency of the column is obtained by a technique based on the measurement of the hydrogen isotope distribution in a steady front attained in a rather long column. In this case, determination of HTU is performed by a technique using stepwise change of the isotope concentration at the column input and measurement of the front smearing at the output [4.9, 10]. Experiments are carried out in a stainless-steel apparatus which is shown schematically in Fig. 4.3.

The basic elements are: the column filled with sorbent (4), gas siphons (1), (2) for hydrogen and for a mixture of hydrogen isotopes equipped with pressure regulators, a reference manometer (5) for measuring working pressures, a gas container (6) for sampling, a meter (7) for hydrogen flow measurements. Granulated

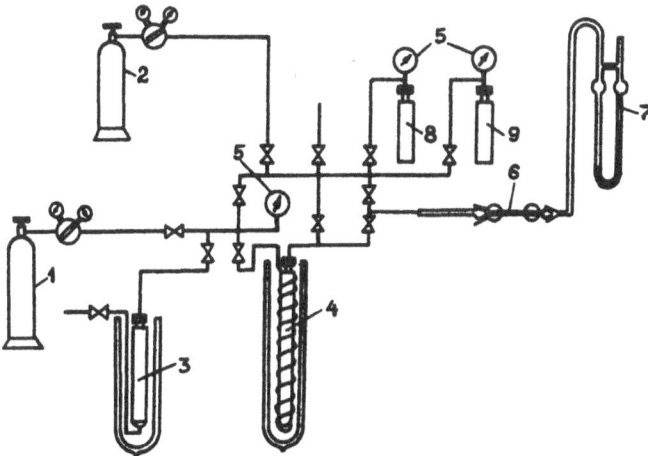

Fig. 4.3. Scheme of plant for the investigation of the rate of hydrogen interphase isotope exchange. (1, 2) compressed gases, (3) trap with zeolite NaX, (4) column, (5) manometers, (6) moveable gas sample container, (7) device for hydrogen flow measurement, (8, 9) hydrogen accumulators

sorbent on the basis of Me or IMC is loaded into the column, which is then evacuated and thermostated. Then a mixture of definite isotope composition is leaked into the column and held until the system attains phase and isotope equilibrium (the required time is estimated on the basis of preliminary experiments on isotope exchange) between the gas and the sorbent in the column. A mixture of another isotope composition is leaked into the system from the gas siphon (2) up to the operating pressure in the column; thereafter the column is connected with the equipment controlling hydrogen flow by means of a valve at the output. The sampling is performed at regular intervals by means of the gas container (6).

From an analysis of the results we plotted the curve of isotope concentration y in the column output as a function of time τ (curve 1, Fig. 4.4), which transforms to the dependence $n^* = f(\tau)$ (curve 2, Fig. 4.4) by the equation

$$n^* = \frac{y - y'}{y'' - y'}, \tag{4.28}$$

where y' is the isotope concentration in the gas forming the hydride phase in the column; y'' is the isotope concentration in the gas blowing through the column; y is the current isotope concentration in the gas at the column output.

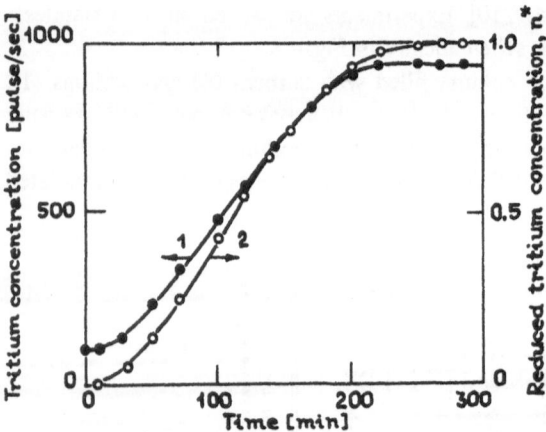

Fig. 4.4. (1) Experimental curves and (2) output curves transformed by (4.28)

The value of the derivative $(\partial n^* / \partial \tau)_{n^*=0.5}$ at $n = 0.5$ is found by graphical differentiation of the curve $n^* = f(\tau)$. The derivative is used for determining HTU from the equations suggested in [4.9]:

$$\frac{1}{u}\left(\frac{\partial n^*}{\partial \tau}\right)_{z,n^*=0.5} = \frac{|\varepsilon^*|}{4h_{0g}}\left(\frac{\exp(-A)}{(n^*A)^{1/2}\mathrm{erfc}(A^{1/2})} - 1\right) \text{ at } \varepsilon^* < 0, \tag{4.29}$$

$$\frac{1}{u}\left(\frac{\partial n^*}{\partial \tau}\right)_{z,n^*=0.5} = \frac{\varepsilon^*}{4h_{0g}}\left(\frac{\exp(-A)}{(n^*A)^{1/2}\mathrm{erfc}(-A^{1/2})} + 1\right) \text{ at } \varepsilon^* > 0. \tag{4.30}$$

The values A, ε^* and u in (4.29, 30) are defined by the equations

$$A = \frac{\varepsilon^* \gamma_{\mathrm{m}} Z}{4h_{0\mathrm{g}}} \; ;$$

$$u = \gamma_{\mathrm{c}} w \; ;$$

$$\varepsilon^* = \frac{(\alpha - 1)(y'' - y')}{1 + (\alpha - 1)y'} \; ,$$

where α is the separation factor, Z the height of the sorbent bed in the column and w the linear gas velocity in the column.

The value $\gamma_{\mathrm{m}} = 1 - \gamma_{\mathrm{c}}$ is calculated by means of the relation

$$\gamma_{\mathrm{c}} = \frac{S_0 c_0 (y'' - y')}{E_{\mathrm{H}}(x'' - x') + S_0 c_0 (y'' - y')} \; , \tag{4.31}$$

where S_0 is the fraction of free section of the column; x' is the initial concentration of the desired isotope in the solid phase; x'' is its concentration in the "waste" sorbent bed; E_{H} is the hydrogen content per unit volume of sorbent (the number of hydrogen moles per unit of sorbent volume); $c_0 = p/RT$ is the number of hydrogen moles per unit volume of gas in the column; T is the temperature in the column; R is the gas constant.

To simplify the determination of $h_{0\mathrm{g}}$ we plotted the dependence $1/u \, (\partial n^*/\partial \tau)_{n^* = 0.5} = f(h_{0\mathrm{g}})$ (Fig. 4.5) using the table presented in [4.11].

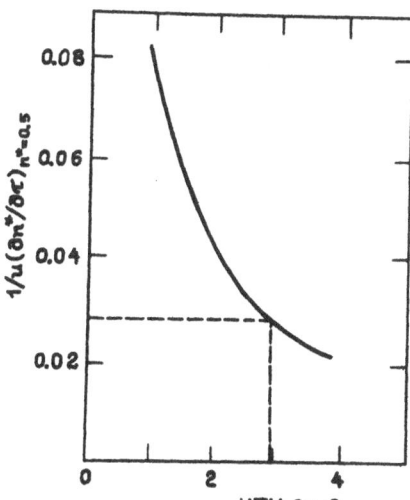

Fig. 4.5. Example of HTU determination by the relationship $1/u \, (\partial n^*/\partial \tau)_{n^* = 0.5} = f(h)$

The use of the suggested technique supposes the following conditions:

$$|\varepsilon^*| \ll 1 \; ;$$

$$Z > \frac{0.04 h_{0\mathrm{g}}}{\varepsilon^{*2} \gamma_{\mathrm{m}}} \; .$$

To fulfill these conditions, the experiments are carried out in the range of tritium trace amounts or using hydrogen containing no more than 15 at% of deuterium.

The values of HTU determined by the above technique relate to the interphase isotope exchange efficiency under conditions of an immobile solid phase. Since the contribution of the component h_s to h_{0g} depends on the flow ratio λ and all the components in (4.20) depend on the loading in the column, let us analyze the loading and flow ratio to which the values of HTU computed from the experimental data correspond. Let us first consider the differences between the velocities of the gas blowing through the column and the effective flow of the gas phase with reference to the moving exchange front affecting the experimental HTU components. The velocity of the exchange front by (4.31) is expressed by the relation

$$u - w = \frac{S_0 c_0 (y'' - y')}{S_0 c_0 (y'' - y') + E_H (x'' - x')} . \tag{4.32}$$

Since $E_H > S_0 c_0$, we have $u \ll w$. In the case of experiments in the range of low heavy isotope content

$$u - w = \frac{S_0 c_0}{S_0 c_0 + \alpha E_H} . \tag{4.33}$$

For example, for zeolite at $P = 0.1\,\mathrm{MPa}$, $T = 77\,\mathrm{K}$, $\alpha \approx 2.0$, $S_0 = 0.48$, $E_H = 276\,\mathrm{mol/m^3}$ and $u/w = 0.015$, i. e., $u \ll w$, the difference in the gas flow velocities with reference to the column walls (w) and to the moving exchange front (u), $u - w$ can be neglected.

Since the specific flows of hydrogen and of solid phase with reference to the moving front are expressed as

$$G'_{sp} = (w - u) S_0 c_0 \tag{4.34}$$

$$L'_{sp} = u E_H , \tag{4.36}$$

from (4.32–34) one can find the following expression for the flow ratio

$$\lambda = \frac{G'_{sp}}{L'_{sp}} = \frac{x'' - x'}{y'' - y'} . \tag{4.36}$$

Thus λ depends on the difference between the isotope concentration in the gas which is pre-sorbed (y') and in the gas which is blown through the sorbent (y''), and on the separation factor, α, on which the concentrations x'' and x' depend. The following equality is valid in the range of low concentrations of the desired isotope: $x'/y' = x''/y'' = \alpha$, and hence $\lambda = \alpha$.

If the mixture of hydrogen isotopes of natural composition initially fills the column or if it is washed out, $\lambda = x''/y''$ (at $x' = y' = 0$) or $\lambda = x'/y'$ (at $x'' = y'' = 0$). That is why when the total expression for the separation factor $\alpha = x(1 - y)/(y(1 - x))$ is required, the flow ratio appears to be slightly lower $(1 < \lambda < \alpha)$.

It is evident that when the heavy isotope is concentrated in the gas phase, the value of λ is below 1.

The efficiency of the mass-transfer in a column, apart from HTU for the gas phase, can also be specified by HTU for the solid phase ($h_{0s} = h_{0g}\alpha/\lambda$), by the value of the height equivalent for the theoretical plate of separation (HETP), and by the mass-transfer coefficients for the gas (K_{0g}) and solid phase (K_{0s}). The relation between HETP (h_e) and HTU for the gas phase is expressed as follows: (a) in the range of low heavy isotope content by

$$h_e = \frac{h_{0g}\ln(\alpha/\lambda)}{1 - \lambda/\alpha};\tag{4.37}$$

and (b) in the range of high heavy isotope content by

$$h_e = \frac{h_{0g}\ln(\alpha\lambda)}{\lambda\alpha - 1}.\tag{4.38}$$

When investigating the mass-transfer efficiency in gas–solid-phase systems, the mass-transfer coefficient for the gas phase is usually used: $K_{0g} = G_{sp}/h_{0g}a_{gr}$.

In parallel with the listed characteristics of mass-transfer efficiency, the volume mass-transfer coefficient (this value reflects both loading and HTU) provides useful informations:

$$K_{0g,v} = K_{0g}a_{gr} = \frac{G_{sp}}{h_{0g}}.\tag{4.39}$$

4.4 Dependence of the Isotope Exchange Rate on the Number of Hydrogenation–Dehydrogenation Cycles During the Metal and IMC Activation Process

Intermetallic compounds produced from powdered or granulated source components require preliminary activation to convert them to a highly active state in which they are easily capable of sorbing and desorbing hydrogen. During the activation one observes a progressive decrease of the IMC particle size and a rise of specific surface up to some constant magnitude.

Consideration of the solid phase dispersity is important both for the comparison of the properties of different heterogeneous systems and in studies of any separate heterogeneous process. The change of dispersity during the preliminary activation reveals the relation between the particle size or specific surface of the powder and the kinetic parameters of the processes in the hydrogen–IMC system. At the same time in the various works on the kinetics of interphase chemical exchange in hydrogen–IMC-hydride systems [4.12–14], a comparison of the kinetics of the process with the data on IMC dispersity is entirely absent.

The influence of the change in IMC dispersity due to activation on the kinetics of the isotope-exchange reaction H–T is shown experimentally in [4.15] for the example of LaNi$_5$ and IMC of composition Ce$_{0.05}$La$_{0.95}$Al$_{0.02}$Ni$_{4.98}$. The experimental data for different degrees of preliminary activation are presented in Fig. 4.6 in coordinates corresponding (4.12).

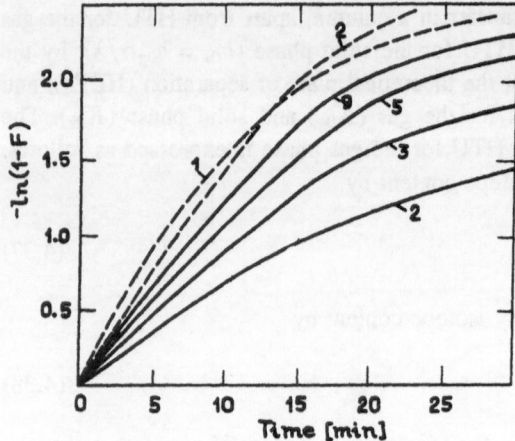

Fig. 4.6. Kinetic curves of hydrogen isotope exchange with LaNi$_5$ hydride (—) after (2), (3), (5), and (9) sorption acts and with Ce$_{0.05}$La$_{0.95}$Al$_{0.02}$Ni$_{4.98}$ hydride (- - -) after (1) and (2) sorption acts at $T = 185$ K and $P = 0.2$ MPa (curve numbers correspond to the number of sorption acts)

As is evident from Fig. 4.6, the curves deviate significantly from straight lines; however, the value of exchange degree for which (4.12) is valid increases with increasing number of sorption acts. A faster stabilization of the shape of the kinetic curves for the sample Ce$_{0.05}$La$_{0.95}$Al$_{0.02}$Ni$_{4.98}$ in comparison to LaNi$_5$ agrees well with the different character of specific surface change as shown in Fig. 4.7. The limiting value of S_{sp} for LaNi$_5$ is only reached after 8–9 activation cycles, but for the sample Ce$_{0.05}$La$_{0.95}$Al$_{0.02}$Ni$_{4.98}$ after 1–2 cycles. It was found that an increase of IMC surface at activation occurs only during sorption acts, since the values of S_{sp} found for hydrides coincide with the corresponding ones for dehydrogenated compound for equal numbers of sorption acts. As is evident from the results of granulometric analysis (Table 4.1), stabilization of the distribution of LaNi$_5$ particles by size also occurs after ~ 8 activation cycles.

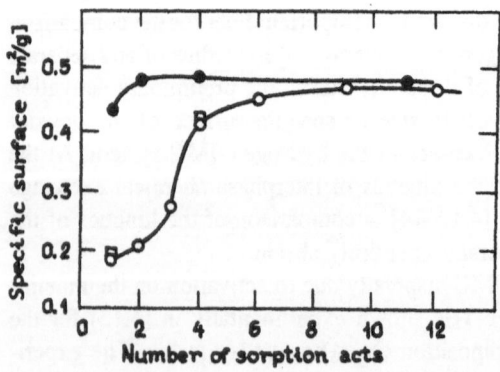

Fig. 4.7. Dependence of the specific surface of IMC on the number of sorption acts:
o LaNi$_5$, □ LaNi$_5$H$_{6.6}$,
● Ce$_{0.05}$La$_{0.95}$Al$_{0.02}$Ni$_{4.98}$

Table 4.1. The results of granulometric analysis of LaNi$_5$ particles (σ is the standard deviation, l_g is the average geometrical size of particles)

The number of activation cycles (T_{sorb} = 185 K)	σ	l_g [μm]
2	6.6	0.52
3	6.4	0.42
4	5.8	0.39
∞	5.6	0.33

The limiting values of the particles sizes and specific surface are found to depend strongly on the temperature of hydrogenation at the activation. Thus for the activation temperatures 293 and 185 K, the difference in the limiting values of S_{sp} is more than 70 % (Table 4.2). This is true because the limiting values of particles sizes and, hence, S_{sp} are determined, all other factors being equal, by the value of the relative expansion of the crystal IMC volume during hydrogenation and by the limit of elastic deformation of the IMC material. The volume expansion increases, and the elasticity limit decreases with decreasing temperature, resulting in a decrease of the particles' limiting size and an increase of the limiting value of S_{sp}.

Table 4.2. Dependence of the limiting value of S_{sp} on the temperature of hydrogenation at activation

IMC	S_{sp} [m^2/g]	
	293 K	195 K
LaNi$_5$	0.27	0.48
CLAN(CeLaAlNi)	0.29	0.49

Based on the values found for S_{sp}, specific constants of the isotope exchange rate R_{sp} for IMC samples with different activation degree (Fig. 4.8b) are computed from the corresponding value of the rate constant R (Fig. 4.8a). Calculation of R is performed using (4.17).

The observed rate constant r is determined by the slope of the initial linear portion of the kinetic curves.

The separation factors for LaNi$_5$ and Ce$_{0.05}$La$_{0.95}$Al$_{0.02}$Ni$_{4.98}$ are found to be equal; furthermore they do not depend on the number of activation cycles and agree with the data known from the literature for LaNi$_5$ (Chap. 3).

As is evident from Fig. 4.8b, an appreciable decrease of the specific rate constant R_{sp} occurs during the preliminary activation of LaNi$_5$ and Ce$_{0.05}$La$_{0.95}$Al$_{0.02}$Ni$_{4.98}$. For the latter, the constant R also decreases because of the slight change of S_{sp} with increasing number of sorption acts (Fig. 4.8a).

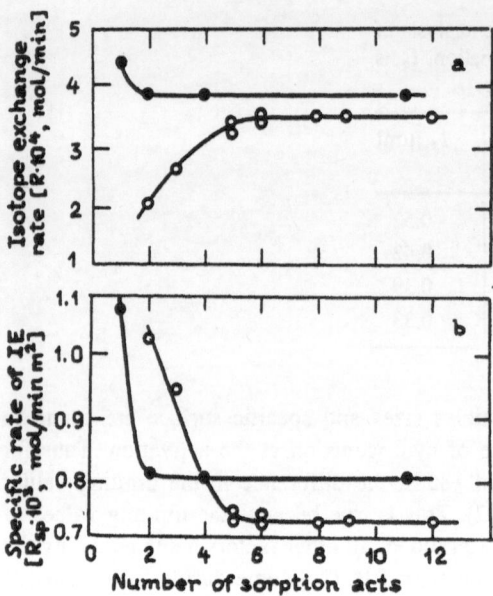

Fig. 4.8a, b. Dependence of the rate constant R (a) and specific rate constant of isotope exchange R_{sp} (b) at $T = 185$ K and $P = 0.2$ MPa on the number of sorption acts: o LaNi$_5$, • Ce$_{0.05}$La$_{0.95}$Al$_{0.02}$Ni$_{4.98}$

Consider in this connection, the published data on the change of the surface structure of IMC of LaNi$_5$-type during the preliminary activation. *Schlapbach* and co-workers [4.16–18], on the basics of a study of LaNi$_5$ surfaces by measurements of magnetic susceptibility and methods of electron spectroscopy, showed that strong segregation occurs for this IMC, increasing with the number of sorption acts. As a consequence, for extended LaNi$_5$ activation, the final surface atomic ratio La : Ni increases to 1 : 1. Segregation is inherent in the majority of IMC and alloys, since it is caused by the difference in the components' surface energy. The tendency of any system towards the energy minimum leads to an enrichment of the surface with the component of lower surface energy (or surface strain). The greater the strain, the higher S_{sp} (or the smaller the particles). That is, the rise of S_{sp} (or decrease of average particle size) results in a decrease in the surface concentration of the metal with high surface energy (it is Ni for the studied IMC). Segregation on LaNi$_5$ at low pressures occurs with sufficient rate even at 18–25 °C [4.16]. One can assume that the surface concentrations of La and Ni corresponding to the particle size are reached in IMC activation in each desorption act (evacuation at elevated temperature).

According to *Schlapbach* [4.18], Ni on the surface of LaNi$_5$ is in the form of metallic particles containing up to 6000 atoms. Exactly these particles are responsible for the catalytic activity on the LaNi$_5$ surface. It is evident that all surface steps of the isotope exchange proceed on these particles, namely, formation (decay) of activated complexes; diffusion of atomic hydrogen on the surface; and injection of hydrogen into the crystal lattice [outcrops]. Since the decrease in the fraction of Ni on the IMC surface during activation is followed by a fall in R_{sp}, one may assume that the rate-determining step of the exchange is related to the IMC surface.

4.5 Influence of Polydispersity of Metal and IMC Samples on the Kinetics of Hydrogen Isotope Exchange

According to the relations of formal kinetics considered in Sect. 4.1 exchange in the range of low concentrations of one of the exchanging isotopes is to be expressed by a first-order equation (4.12). However, as is noted above, the kinetic curves presented in Fig. 4.6, deviate significantly from straight lines, despite referring to isotope exchange in the range of tritium trace amounts.

Let us show that if surface processes are the rate-determining factor, solid phase polydispersity is the cause of this deviation (even if R_{sp} is constant and does not depend on the particle size). It can also be shown that the above-mentioned dependence of R_{sp} (as well as polydispersity itself) on S_{sp} or the particle size also result in deviation from the simple exponential law.

Let us introduce the following symbols: l is the average size (by volume) of IMC particles; $F(l)$ is exchange degree for particles of size l; $\varphi(l)$ is the mass accounted for by these particles as a fraction of the mass of the whole sample; n_l is the number of particles of size l in the sample; $g(l)$ is mass of particle of size l; ϱ is IMC density; ε_v, ε_s are the volume and surface form factors; $r(l)$, $R_{sp}(l)$ are the observed and specific rate constants of the isotope exchange for particles of size l; n is the amount of hydrogen per unit volume of solid phase; S, V are the total surface and volume of all particles in the sample.

Let us arbitrarily split the considered heterogeneous system into a number of subsystems, each consisting of a separate particle with a hydrogen amount in the solid phase of $N_\phi^s(l)$ and in the gas phase of $N_\phi^g(l)$: The total exchange degree for the whole system is expressed in terms of its values for the subsystems as follows:

$$F = \sum_l \phi(l) n_l F(l). \tag{4.40}$$

One can extract the variable l involved in $\phi(l)$ and $r(l)$, albeit in unexplicit form:

$$\phi(l) = \frac{g(l)}{g} = \frac{\varrho \varepsilon_v l^3}{g} = \frac{\varepsilon_v l^3}{V}. \tag{4.41}$$

For each separate subsystem we have:

$$r(l) = R_{sp}(l) \varepsilon_s l^2 \frac{\alpha n^s \varepsilon_v l^3 + N_\phi^g(l)}{\alpha n^s \varepsilon_v l^3 + N_\phi^s(l)}. \tag{4.42}$$

Let us introduce the notation

$$Z(l) = \varepsilon_s l^2 \frac{\alpha n^s \varepsilon_v l^3 + N_\phi^g(l)}{\alpha n^s \varepsilon_v l^3 N_\phi^g(l)} = \frac{\varepsilon_s}{l \varepsilon_v} \left(\frac{V}{N^g} + \frac{1}{\alpha N^s} \right) = \frac{\xi}{l} \left(\frac{N^s \alpha + N^g}{N^g n^s \alpha} \right) \tag{4.43}$$

where $\xi = \varepsilon_s / \varepsilon_v$. This yields

$$r(l) = R_{sp}(l) Z(l). \tag{4.44}$$

The following first-order equation holds for each separate subsystem

$$F(l) = 1 - \exp[-r(l)\tau] = 1 - \exp[-R_{sp}(l)Z(l)\tau].$$ (4.45)

This relation can be written for the whole sample, in view of (4.40, 41, 44), as:

$$F = \frac{\varrho \varepsilon_v}{g} \sum_l n_l l^3 \{1 - \exp[-R_{sp}(l)Z(l)\tau]\}$$

$$= \frac{\varrho \varepsilon_v}{g} \sum_l n_l l^3 - \frac{\varrho \varepsilon_v}{g} \sum_l n_l l^3 \exp[-R_{sp}(l)Z(l)\tau].$$ (4.46)

Since

$$\frac{\varrho \varepsilon_v}{g} \sum_l n_l l^3 = \sum_l \phi(l) n_l = 1,$$ (4.47)

we obtain the final kinetic equation for a polydispersed sample:

$$F = 1 - \frac{\varrho \varepsilon_v}{g} \sum_l n_l l^3 \exp[-R_{sp}(l)Z(l)\tau].$$ (4.48)

For a known character of the granulometric distribution, the parameter n_l can be replaced by the function describing the distribution of particles by size $n(l)$, in view of the normalizing condition similar to (4.47):

$$\sum_l \phi(l) n(l) = 1,$$ (4.49)

i. e.,

$$F = 1 - \frac{\varrho \varepsilon_v}{g} \sum_l n_l l^3 \exp[-R_{sp}(l)Z(l)\tau].$$ (4.50)

For an infinitely large number of particles in the IMC sample, the discrete variable l becomes continuous, whence

$$F = 1 - \frac{\varrho \varepsilon_v}{g} \int_{l_{min}}^{l_{max}} n(l) l^3 \exp[-R_{sp}(l)Z(l)\tau].$$ (4.51)

It is evident that, for a monodispersed character of IMC powder, equations (4.48, 50, 51) transform to (4.12). In the presence of a distribution of particle sizes, one has, for small τ only,

$$\exp[-R_{sp}(l)Z(l)\tau] \approx 1 - R_{sp}(l)Z(l)\tau,$$ (4.52)

$$F = 1 - \frac{\varrho \varepsilon_v}{g} \sum_l n_l l^3 [1 - R_{sp}(l)Z(l)\tau].$$ (4.53)

For infinitesimal τ and using the normalizing condition (4.49) we obtain the first-order equation

$$-\ln(1 - F) = \tau \sum_l \phi(l) n(l) R_{sp}(l) Z(l),$$ (4.54)

where

$$r = \sum_l \varphi(l)n(l)R_{sp}(l)Z(l). \tag{4.55}$$

Equation (4.55), together with (4.41) and (4.43), gives

$$r = \frac{1}{N_g} \sum \varepsilon_s l^2 n(l) R_{sp}(l) + \frac{1}{\alpha n^s V} \sum \varepsilon_s l^2 n(l) R_{sp}(l)$$

$$= \frac{SR_{sp}}{N_g} + \frac{SR_{sp}}{\alpha N_s} = Rg\left(\frac{1}{N_g} + \frac{1}{\alpha N_s}\right). \tag{4.56}$$

This corresponds to (4.12). Thus (4.12) is also true for polydispersed samples if r is replaced by the value of the slope of the initial linear portion of the curves as is done in the work presented below. In this case the constant r relates to the overall geometrical surface of an IMC sample and R_{sp} is a value averaged over all particle sizes.

As τ increases the kinetic curves increasingly deviate from first order. Since in the limiting case of monodispersed powder this deviation is not observed at any τ, the lower the polydispersity, the slighter the deviation of the kinetic curve from (4.12) at the same τ. As it is obvious from Table 4.1, the standard deviation σ characterizing the polydispersity decreases with increasing number of activation cycles. It agrees well with the change in form of the kinetic curves during the activation; the length of the initial linear portion in coordinates $-\ln(1-F)$ vs τ increases with increasing number of sorption acts and a deviation from the linear form is observed at higher values of exchange degree F.

Deviation of the kinetic curve from first order is caused by the difference in $r(l)$ for particles of different size. Decrease of l results in the rise of $r(l)$ due to the greater specific surface of the particles but simultaneously leads to some decrease of $R_{sp}(l)$ due to enhanced segregation. That is, in the presence of segregation, the difference in $r(l)$ for different particles is smaller than for constant R_{sp} and, hence, the initial linear portion of the kinetic curves is longer.

Consider now the other limiting case in which τ tends to infinity. As is evident from (4.43), the function $Z(l)$ is proportional to $1/l$. The function $R_{sp}(l)$, as is shown below, is correlated with l by (4.64). An analysis performed in terms of the form of dependences of $R_{sp}(l)$ and $Z(l)$ on l shows that the value of the exponent in (4.50) increases from 0 to 1 as l changes from 0 to infinity. Therefore, the last term in (5.50) corresponding to $l = l_{max}$ becomes greater than the remaining terms with increasing τ. Equation (4.57) appears to be the exponential asymptote of (4.50) at $\tau \to \infty$:

$$F = 1 - \varphi(l_{max})n(l_{max}) \exp\left(-R_{sp}(l_{max})Z(l_{max})\tau\right), \tag{4.57}$$

or, in logarithmic form,

$$\ln(\varphi(l_{max})n(l_{max})) - \ln(1 - F) = r(l_{max})\tau. \tag{4.58}$$

In the plot of $-\ln(1 - F)$ against τ the line given by (4.58) intersects the ordinate at a point that corresponds to the mass fraction of the largest particles in a given sample. If l_{max} is known, it is assumed to be possible to calculate the value of the specific constant corresponding to the given particle size:

$$R_{sp}(l_{max}) = \frac{\tau(l_{max})\alpha N_g N_s l}{\xi(\alpha N_g + N_s)} . \tag{4.59}$$

This constant appears to be the "true" value, i.e., it is related to a definite size of particles as opposed to R_{sp} which is averaged over all particles. However, in practice, because of the abrupt rise of the error in determination of F as equilibrium is approached, such a calculation is possible only when the largest particles constitute a considerable part of the overall sample.

For theoretical calculation of the kinetic curves using (4.51), in addition to the data determined by experimental conditions and the chosen system (N_g, N_s, α, g, ϱ), one also needs to know the exact form of the functions $n(l)$ and $R_{sp}(l)$ and the ratios of the form coefficients ξ. As is evident from granulometric investigations, the distribution of IMC particles by sizes is well described by a normal logarithmic distribution:

$$f(\ln l) = \frac{1}{\ln \alpha (2\pi)^{1/2}} \exp\left[\frac{-(\ln l - \ln l_g)^2}{2\ln^2 \sigma}\right], \tag{4.60}$$

where $\sigma = [\sum(\ln l - \ln l_g)^2 n/N]^{1/2}$ is the standard deviation and N is the total number of particles. The distribution (4.60) is related to the function $n(l)$ by the relation

$$f(\ln l) = Bn(l), \tag{4.61}$$

where B is a coefficient determined by the normalizing conditions. The distribution parameters l_g and σ are computed for the powder from the experimental data and are summarized in Table 4.1.

Finding the form of $R_{sp}(l)$ is a more difficult task. The experimental data (Fig 4.8) give only the relation of the averaged constant R_{sp} to S_{sp} (being proportional to $1/l$). The averaged R_{sp} is connected with the constants $R_{sp}(l)$ for separate particles by the expression

$$R_{sp} = \sum_l R_{sp}(l)\phi(l)n(l), \tag{4.62}$$

It is evident that the dependence $R_{sp}(l)$ is different to the experimental one. As noted above, $R_{sp}(l)$ for each particle is proportional to the fraction of Ni on its surface X_{Ni}^S. The latter is related to the Ni atomic fraction in the bulk X_{Ni}^b by the following equation [4.19]

$$\frac{X_{Ni}^S}{X_{La}^S} = \frac{X_{Ni}^b}{X_{La}^b} \exp\left(\frac{a(\sigma_{La} - \sigma_{Ni})}{RT}\right), \tag{4.63}$$

where σ_{La} and σ_{Ni} are surface strains and a is the surface area per mole.

Considering that $a \propto 1/l$ and $\sigma_{La} < \sigma_{Ni}$ we obtain

$$R_{sp}(l) = A \exp\left(\frac{-C}{l}\right)$$

where A and C are constants independent of l.

However, one must remember that (4.63) is applicable only when a surface monolayer is considered and when the complicating chemical influences of hydrogen impurities on the segregation are neglected.

The ratio of the form coefficients ξ can be determined by comparison of S_{sp} values computed from the granulometric distribution with the experimental values. For the IMC studied, the values of ξ are found to be in the range 8–9.

Thus, when the functions $n(l)$ and $R_{sp}(l)$ are known, the kinetic curves of IIE for real polydispersed samples can be computed using (4.51). For systems, in which R_{sp} does not depend on the particle size, the calculation is much simpler. Analysis of the experimental kinetic curves gives information on the change of dispersion characteristics and properties of the IMC surface during the sample activation.

4.6 Influence of Temperature and Pressure on Isotope Exchange Kinetics

Investigation of the influence of temperature and pressure on the isotope exchange rate in hydrogen-isotopes–metal/IMC-hydride systems enables one to establish the rate-determining step of the process and to find ways to accelerate it. The majority of the experimental data on the influence of temperature and pressure has been obtained by the present authors and their co-workers for conditions $\alpha \neq 1$ and $x, y \ll 1$, corresponding to (4.12, 17). In general form, since hydrogen pressure determines its concentration in the solid phase, (4.1) can be represented as

$$R = R_0 \exp\left(\frac{-E}{RT}\right) P^m, \tag{4.64}$$

where E is the observed activation energy of isotope exchange and m is the order of the isotope exchange reaction with respect to hydrogen pressure.

The apparent activation energy of the isotope exchange reaction is found by a graphical method from the results of experiments carried out at different temperatures and constant hydrogen pressure in accordance with the equation

$$\ln R = \ln R_0 - \frac{E}{RT}. \tag{4.65}$$

When carrying out the experiments at constant temperature and different hydrogen pressures, the order of reaction can be found from

$$m = \frac{\Delta \ln R}{\Delta \ln P}. \tag{4.66}$$

Figure 4.9 presents the temperature dependence of the rate of hydrogen isotope exchange for a number of IMC hydrides and Fig. 4.10 shows the dependence of R on hydrogen pressure. The thermodynamic and kinetic characteristics of isotope exchange of hydrogen with hydrides of a number of metals and IMC are summarized in Table 4.3.

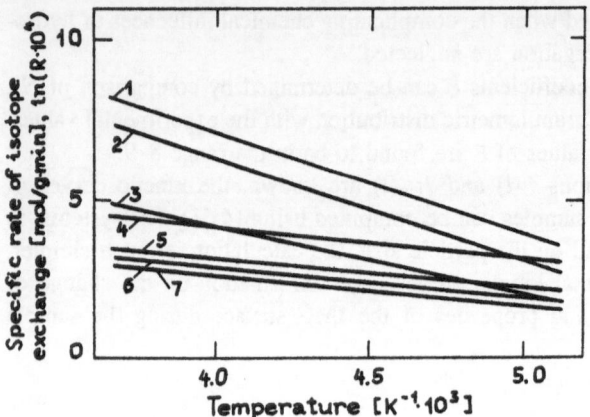

Fig. 4.9. Dependence of the isotope exchange rate on temperature: (1) LaNi$_4$Cr, (2) LaNi$_4$Cu, (3) LaNi$_5$, (4) TiCrMn, (5) TiMn$_{1.4}$Ni$_{0.1}$, (6) Ti$_{0.8}$Zr$_{0.2}$Cr$_{1.8}$, (7) TiMn$_{1.5}$ at $P = 0.4$ MPa

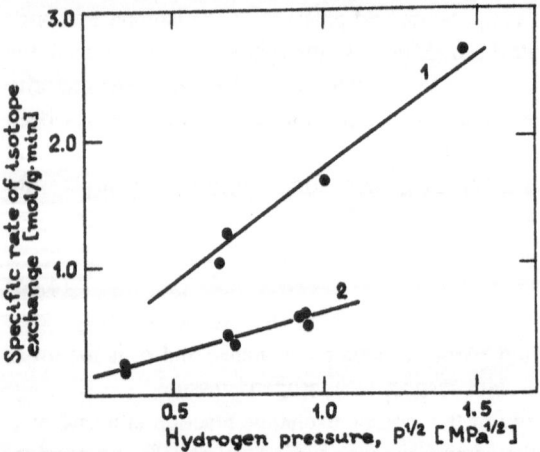

Fig. 4.10. Dependence of the isotope exchange rate on hydrogen pressure: (1) TiCrMn, (2) TiMn$_{1.5}$

It should be noted that for all activated samples of metals and IMC over a wide range of temperatures and pressures the first-order kinetic equation (4.17) remains valid up to high degrees of exchange ($F \geq 0.8$), and enables one to determine the R value characterizing the averaged isotope exchange rate for particles of different size from the slope of the straight line in coordinates $-\ln(1 - F) = f(\tau)$. Before proceeding to the mechanism of the isotope exchange reaction and analysis of the results shown in Table 4.3 we dwell on the experimental results obtained by different authors in studies of isotope exchange in the hydrogen–palladium hydride system by the static method with forced gas circulation.

Table 4.3. Thermodynamic and kinetic characteristics of interphase isotope exchange (IIE) of hydrogen with IMC hydrides at $P_{H_2} = 0.5\,MPa$

IMC hydride	α_{HT} at 195 K	$R_{HT} \times 10^{-4}$ mol/(g · min) at 195 K	E [kJ/mol]	$\ln R_0$	m
$LaNi_5H_{6.6}$	2.12	6.5	17	3.4	0.3
$LaNi_4CuH_{5.5}$	2.12	1.4	28	11.5	0.4
$LaNi_4CrH_{5.0}$	2.12	2.6	27	10.1	0.4
$TiMn_{1.5}H_{2.5}$	3.10	4.1	9.4	−2.0	0.5
$TiMn_{1.4}Ni_{0.1}H_{2.3}$	2.80	8.1	8.5	−1.9	0.5
$TiCrMnH_{0.3}$	3.20	2.1	8.5	−0.9	0.5
$Ti_{0.8}Zr_{0.2}CrMnH_{2.9}$	3.20	2.1	8.5	−0.9	0.5
$Ti_{0.8}Zr_{0.2}Cr_{1.8}H_{2.8}$	3.17	5.3	9.1	−1.9	0.4
$ZrMn_2H_3$	1.75*	27.1*	15.0	0.7	0.4
$ZrCr_2H_3$	2.32*	10.6*	17	0.64	0.5
$ZrV_2H_{4.3}$	1.77*	13.3*	12	−1.2	0.5

* data at 273 K

The study [4.24] is the first in which, along with isotope equilibrium in the system $H_2(D_2)$–Pd, the kinetics of isotope exchange on granulated palladium powder (2–3 mm) is investigated.

In the temperature range 273–363 K and the pressure range 46–532 mbar, isotope exchange is expressed by (4.64) with the following coefficients: $E = 30.4\,kJ/mol$ and $m = 1$. The experiments on the dependence of the exchange rate on pressure are carried out only at a temperature of 303 K. The authors of [4.24] assumed that at low temperatures, change of the isotope exchange mechanism is possible. This is corroborated by more recent works [4.13, 25], in which isotope exchange on palladium powder and on palladium applied to aluminum oxide was studied. The experimental data from [4.13, 25] are presented in Table 4.4.

Based on the data obtained, especially the value of the activation energy, one can conclude that the rate-determining step of the isotope exchange process at low temperatures is transport of hydrogen atoms from the surface to interstitial sites. In the case of $Pd/\gamma Al_2O_3$ diffusion of molecular hydrogen in the pores of the carrier is the rate-determining step [4.25]; The influence of diffusion in the gas phase is excluded due to the high-rate of hydrogen circulation through the sorbent. Note that in all the works considered hydrogen is initially purified by low-temperature sorption of nitrogen on activated carbon, of oxygen on reduced Cr–Ni catalyst, and

Table 4.4. Rates and activation energy of hydrogen isotope exchange on palladium sorbents

Sorbent	d_m nm	S_{SP} m²/g	T K	$R_{H-D} \times 10^7$ mol/m²s	E kJ/mol
Pd powder	40	10.2	167	0.6	20.5
			175	1.2	
	200	2.5	167	0.7	
			175	1.3	
			195	6.1	21.0
Pd/γAl$_2$O$_3$	4	102	195	3.0	
			228	3.0	< 2.6
			296	3.0	
Pd/$\varkappa$Al$_2$O$_3$	50	7.4	175	1.2	
			195	4.7	18.9

of moisture on silicagel and zeolites at liquid nitrogen temperature. The presence of water impurity is checked by means of an isotope composition measurement.

An investigation of the kinetics of isotope exchange on palladium of hydrogen labeled with tritium is reported in [4.14]. Isotope exchange was studied in the temperature range 247–323 K at a hydrogen pressure of 100 mbar; the activation energy is 11 kJ/mol. Such a large difference cannot be related to the substitution of deuterium with tritium. The difference can be correlated with the purity of the hydrogen used in the experiments; in [4.14] extremely pure hydrogen is used (of purity 5.0). Subsequent works [4.26, 27] corroborated the influence of hydrogen purity on the activation energy of the isotope exchange reaction. Let us dwell on these studies more comprehensively, since the flowing method combined with laser Raman spectroscopy analysis is used in them. This method of analysis enables one to determine the partial pressure of isotope modifications of the hydrogen molecule including those containing tritium.

A comprehensive description of the apparatus and experimental technique is presented in [4.28]. The authors concluded that under experimental conditions ($P_{H_2(D_2)}$ = 1396–1596 mbar, T = 300 K, G = 0.6–1.1 ls.t.p./min) the isotope exchange is determined by surface processes. Thereafter they suggested a model allowing an adequate description of the ratio of H$_2$, HD, and D$_2$ at the output of the reactor. Using a similar technique *Carstens* et al. [4.26, 27] showed that at $P_{H_2(D_2)}$ = 7 bar in the temperature range 173–299 K at a flow G = 2 ls.t.p./min and temperatures higher than 210 K the isotope exchange is determined by diffusion in the gas (on condition that it is highly pure). At lower temperatures the surface processes become rate-determining, which is corroborated by the influence of impurities CH$_4$, CO$_2$, H$_2$O, and CO on the rate of isotope exchange and activation energy.

Table 4.5 shows the results obtained in [4.27] for isotope exchange of hydrogen in the presence of impurities of concentration 100 ppm. The activation energy is determined by the apparent rate constant r – see (4.12) – which depends on the experimental conditions (N_g, N_s, α).

Table 4.5. Activation energy and rate constant for isotope exchange on palladium hydride

Impurity in hydrogen	H–D	exchange	D–H	exchange
	E [kJ/mol]	r [sec^{-1}]	E [kJ/mol]	r [sec^{-1}]
extremely pure	8.8 ± 0.4	7.4	12 ± 1	25
CH$_4$	8.4 ± 0.4	5.1	8.4 ± 1.7	17
CO$_2$	18 ± 0.8	370	22 ± 1.7	8500
H$_2$O	33 ± 3	3.6×10^4	23 ± 5	1800
CO	96 ± 13	2.9×10^{14}	96 ± 16	6.4×10^{15}

Neglect of the temperature dependence of α can result in serious error in determining the activation energy. It is to be noted that when carrying out the experiments on "pure" H–D or D–H exchange, the error of the process simulation can reach 20 % of the isotope exchange rate due to the concentration dependence of α_{H-D}. The difference in the experimental conditions in studies of isotope exchange precludes the possibility of correctly comparing the rate constants with one another; it may simply be noted that these values coincide within an order. The conclusion about the rate-determining step is common to all studies, although there are distinctions in the treatment of the surface processes. For completeness let us analyze the data (obtained with the technique of discrete change of the isotope concentration at the input of the column and measurement of the front smearing at the outlet) obtained on granulated palladium powder and granulated palladium-containing sorbents [4.13, 29]. The influence of temperature, pressure, and loading on the rate of interphase isotope exchange (IIE) of hydrogen is investigated in [4.29] with using isotope mixtures H–D and H–T on granulated palladium powder (grain size 0.3–0.5 mm). The experimental data are analyzed using (4.28–31), yielding HTU or h_{0g}, which is related to the components of the mass-transfer process by (4.20).

Figure 4.11 presents the dependence of HTU on loading for gas at low temperatures. The experiments on the dependence of IIE rate on loading and pressure at high temperatures (293–373 K) showed that independently of pressure and load $h_{0g} \approx 2$ cm, enabling diffusion of the gas towards the sorbent surface to be considered as the rate-determining factor. This conclusion agrees with the data of [4.27]. At low temperatures the activation energy is determined by the dependence $\ln \beta_s a - 1/T$, which is found to be equal to 20.5 kJ/mol for both isotope mixtures. Since this value agrees well with the activation energy of hydrogen diffusion in the β-phase of palladium, we concluded that at low temperatures the

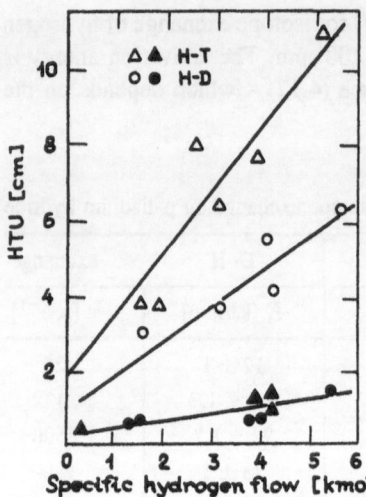

Fig. 4.11. Dependence of HTU (h_{0g}) for H–D and H–T mixtures on specific hydrogen flow G_{sp} in the H_2–Pd system at $P = 0.5$ MPa. △ H–T, o H–D; empty symbols relate to 195 K, filled ones to 273 K

process is determined by hydrogen diffusion in Pd. In the authors' opinion [4.29], the equality $D_D/D_T \approx \beta_s^D/\beta_s^T \approx 1.5$ corroborates this conclusion.

Subsequently, in studies of isotope exchange on grains consisting of Pd-powder and PTFE ($d_{gr} = 1.3$ mm) it is shown [4.13] that at low temperatures the exchange kinetics is determined by the processes of hydrogen atom transport from the surface into interstices of the Pd crystal lattice and at high temperatures the rate is determined by diffusion in pores of the sorbent grains. The data on HMIE on Pd are required to establish the rate-determining step of the isotope exchange at low temperatures both on Pd-powder and on grains.

The HMIE reaction of protium with deuterium is investigated in [4.30] in the temperature range 195–500 K and in the pressure range 1.3–352 mbar. It is shown that HMIE proceeds at a high rate on activated β-Pd-hydride with an activation energy of 9.2 kJ/mol. In this case the rate of the HMIE reaction, even at 195 K, is 3 times higher than the rate of the IIE reaction. Estimation of the half-exchange time for Pd, with hydrogen diffusion in the metal taken as the rate-determining factor, is performed using the equation

$$\tau_{0.5} = \frac{\pi r^2}{144 D_H} \tag{4.67}$$

with the diffusion coefficient of hydrogen atoms at 195 K for the β-phase of Pd being $D_H = 4.2 \times 10^{-13}$ m^2/s [4.31] for the maximal diameter 2×10^{-6} m. The calculation yields the value ≈ 0.1 s, which is distinctly lower than real observable values of $\tau_{0.5}$. This supports the conclusion that it is just hydrogen atom implantation in the crystal lattice of Pd that determines the isotope exchange at low temperatures.

Now we revert to systems with IMC hydrides. Contrary to Pd, data are available on the dependence of the rate constant R on pressure and on HMIE obtained for samples identical to those for IIE.

The kinetics of HMIE of hydrogen on hydrides of $LaNi_5$, $Ti_{0.8}Zr_{0.2}CrMn$, CrMn, $ZrMn_2$ is investigated in [4.20, 23, 32]. The existence of published data on diffusion coefficients of hydrogen atoms in IMC [4.33] and on their particle sizes after activation [4.34–37] enables one to perform a general analysis of the mechanism of isotope exchange between hydrogen and IMC hydrides.

Taking into account that correct performance of the experiments using the static method with forced gas circulation implies that external diffusion deceleration is absent, let us consider the possible rate-determining steps of the IIE process in hydrogen–IMC-hydride systems:

1) formation (decay) of an activated complex or dissociative adsorption (associative desorption) of hydrogen on the IMC surface;
2) hydrogen atom migration on the surface or surface diffusion;
3) conversion from the adsorbed to the absorbed state and vice versa;
4) diffusion of atomic hydrogen in the crystal lattice of IMC.

Step 1 is common to processes HMIE and IIE of hydrogen; the results obtained in [4.20, 23, 32] show that at temperatures up to $100\,K$ a change of the mechanism of the HMIE reaction, which is followed by a change of activation energy, does not occur. The following values are obtained in the temperature range 77–250 K: $E = 6.3, 5.1, 0.9\,kJ/mol$ for $LaNi_5$, $Ti_{0.8}Zr_{0.2}CrMn$, $ZrMn_2$, respectively. At equal temperatures (comparison is performed by extrapolation of the data on HMIE to the range of higher temperatures $\leq 300\,K$ and of the data on IIE to the range of lower temperatures $\geq 100\,K$) the rate of HMIE is significantly higher than the rate of IIE. This indicates that the IIE process cannot be determined by step 1.

Step 4 cannot be the rate-determining factor for the IMC presented in the Table 4.3. Let us corroborate this fact by a calculation using (4.67).

The results of calculating the half-exchange time at low temperatures for a number of IMC are summarized in Table 4.6.

Table 4.6. Estimation of the half-exchange time by diffusion coefficients

IMC hydride	E_{act} kJ/g-atH	$D_H \times 10^{12}$ m^2/sec $T=300\,K$	$D_H \times 10^{14}$ m^2/sec $T=195\,K$	$\tau_{0.5}$ sec $T=300\,K$	$\tau_{0.5}$ sec $T=195\,K$	$d_{Me} \times 10^6$ m
$LaNi_5H_6$	24.0	2.1	1.1	0.3	50	10 [4.34]
	26.5	5.0	1.5	0.001	0.4	1 [4.35]
$TiMn_{1.5}H_3$	21.7	9.8	8.3	0.002	0.3	2 [4.36]
$Ti_{0.8}Zr_{0.2}CrMnH_3$	21.2	6.2	5.8	0.001	0.1	1 [4.37]

As is evident from Table 4.6, the half-exchange time ranges from 0.001 to $50\,s$ depending on the temperature and the size of IMC particles.

Step 2 cannot be the rate-determining factor either since in this case the dependence of the IIE rate R on pressure contradicts that observed (Table 4.3). The rate of surface diffusion is known to be proportional to the gradient of the surface hydrogen concentration itself due to the fact that different regions of the surface are unequal in energy; pressure increase results in filling of the regions that are less energetically favorable and reduces the surface concentration gradient, i. e., decreases the process rate. This does not agree with the established increase of R with increase in pressure. Thus step 3 is the only one that fits the observed kinetic relationships. It is interesting to note that partial substitution of La in LaNi$_5$ with Ce does not result in any essential change of the HMIE rate and also decreases the IIE rate. A similar pattern is also observed for AB$_2$ compounds. It indicates that injection of minor amounts of other elements affecting both the surface properties and the surface layer composition is one way to raise the rate of IIE.

4.7 Hydrogen Isotope Exchange on Granulated Sorbents

4.7.1 Preparation of Granulated Sorbents

It is well known that IMC increase their crystal lattice volume (by up to 25 %) upon the formation of hydrides. This is one of the possible causes of deformation and destruction of construction elements using IMC. Furthermore, formation of fine-grained powders upon IMC activation hinders their practical application due to the particles' entrainment from apparatuses and high gas flow resistance of the powder bed. The preparation of such hydrogen sorbents on the basis of IMC, whose grains offer high resistance to destruction through repeated hydrogenation, thermal and radiation effects (the last is important in terms of separating tritium-containing mixtures), is a topical problem. Many investigations have been devoted to the task of preparing granulated sorbents.

The method of IMC powder shaping with polymeric materials [4.38] is widely used. Polytetrafluoroethylene (PTFE) was the first to be used as a binder [4.39], however, sorbents with PTFE decompose at high temperatures ($\approx$ 450 K) due to IMC interaction with fluorine atoms [4.40]. Besides, PTFE is the least radiation-resistant polymer [4.41]. Polymeric materials, such as polyimides and silicon rubbers, offer reasonably high thermal and radiation resistance [4.42, 43], but the porosity of grains with silicon rubber is found to be moderate and it retards the rate of hydrogen sorption [4.44].

Low thermal conductivity is the common drawback of sorbents that contain polymeric materials. Improvement of heat conduction of granulated sorbents can be achieved by injecting metals that do not form hydrides as a binder component. As is shown in [4.45], porous compositions of IMC powders and metals not forming hydrides satisfy the requirement of fast heat supply and removal from the sorbent bed in cycles of hydrogen sorption–desorption. However, the sorbents prepared by IMC powder pressing in a metal matrix do not withstand large number of cycles of hydrogen sorption–desorption [4.46].

The work [4.47] proposes the interesting preparation technique of sorbents on the basic of IMC being tolerant to destruction by formation of micrograins of

IMC particles covered by a thin copper [or nickel] coat with subsequent pressing. Such a sorbent withstands 1000 sorption–desorption cycles without appreciable degradation.

A widely used method of preparation of sorbents that are resistant to frequently repeated hydrogenation is the following: powders of partially or completely hydrogenated IMC are pressed and baked with powders of metals not forming hydrides, such as Ag, Ni, Cr, Cu, Al, Pb or their alloys [4.46, 48–50]; however, even after 5 cycles decomposition of the sorbents is observed. The modified method of *Ron* et al. [4.46] enables one to attain more sorbent resistance. In this case powder of activated and totally hydrogenated IMC, whose surface has been passivated by means of carbon oxide, is used to prepare the sorbent. After pressing at a pressure of 500 MPa and regeneration by the method described in [4.51] the dehydrogenated sorbent contains a large volume of pores and, hence, hardly changes its overall volume upon hydrogenation. After 50 cycles no decline of the sorption–desorption rate is observed [4.42].

A porous sorbent able to withstand no less than 6000 cycles with no destruction can be prepared by mixing activated IMC powder and a metal not forming a hydride, followed by hydrogenation and subsequent pressing with simultaneous baking at a temperature of 370–470 K in a hydrogen atmosphere at a pressure higher than the pressure of hydride formation. The rate of hydrogen sorption–desorption for such a sorbent is considerably higher than for powdered IMC [4.49].

Comparative analysis of properties of sorbents prepared with the use of various polymeric materials and metals not forming hydrides shows that the granulation method depends on the properties of the IMC and on the intended application of the granulated sorbent. So, for instance, for hydrogen isotope separation, apart from thermal conductivity, the sorbent porosity, which can vary with the rate of thermal treatment during preparation of sorbents with polymeric binders [4.53], is an important characteristic.

4.7.2 Experimental Data on Hydrogen Isotope Exchange on Granulated Sorbents

On granulated sorbents hydrogen diffusion in sorbent pores is added to the steps of the IIE process considered in Sect. 4.6. The influence of this step on the exchange can be revealed by studying the isotope exchange kinetics by the two techniques outlined in Sect. 4.3. Let us analyze the experimental data presented in [4.5, 13, 44, 54]. In these works granulated sorbents with polymeric binder were used. In [4.13] PTFE is applied and in the others polyimide (30 mass%).

Inactivated fine-grained powder (grain size up to 40 μm) of $LaNi_5$ and $Ti_{0.8}Zr_{0.2}CrMn$, which have high thermodynamic (α) and kinetic (R) characteristics (Table 4.3), are used for the preparation of the sorbents.

The sorbents obtained appear to be thermally resistant up to 673 K, i. e., this temperature is high enough for complete hydrogen desorption. It is also important that interaction of hydrogen with the polymeric binder does not occur up to this temperature despite the known catalytic properties of IMC in the hydrogenation and dehydrogenation reaction [4.55].

As is evident from irradiation tests under hydrogen at a pressure of 0.1 MPa and at temperatures of 208–313 K, ^{60}Co γ-radiation does not appreciably affect the sorbent strength (115–121 MPa, approximately twice as high as the zeolite strength). The main characteristics of the sorbent studied are presented in Table 4.7. The grain diameter ranged from 1.0 to 1.5×10^{-3} m; Table 4.7 shows the average values a_{gr} and d_e used for calculation. The sorbent density and the free space were determined experimentally. The specific surface was measured by the method of low-temperature krypton sorption using the BET equation. The average pore diameter was found by means of statistic processing photographs of sorbent sections obtained using scanning electron microscopy.

Table 4.7. Main characteristics of sorbents on the basis of LaNi$_5$ and Ti$_{0.8}$Zr$_{0.2}$CrMn containing 30 mass% of polyimide

Characteristic	Magnitude
Grain diameter [m]	1.25×10^{-3}
Poured sorbent density [g/cm^3]	2.0
Free space, relative units	0.4
Grains surface [m^2/m^3]	2.9×10^3
Equivalent diameter of channels [m]	0.5×10^{-3}
Specific surface [m^2/g IMC]	0.4
Average diameter of pores [μm]	80

The experimental dependence of the IIE efficiency for a H–T mixture at diffrtent temperatures on the specific flow in a column of diameter 1×10^{-2} m and filled with sorbent based on LaNi$_5$ at $P_{H_2} = 0.5$ MPa is presented in Fig. 4.12 and filled with sorbent based on Ti$_{0.8}$Zr$_{0.2}$CrMn at $P_{H_2} = 1.9$ MPa in Fig. 4.13. These figures show that, regardless of temperature, for one or another sorbent h_{0g} depends linearly on the specific molar flow (= molar flow density = load of the column) at $G_{sp} \leq 4$ mol/m^{2}s. All experiments are carried out in a column of length 0.46 m. In addition, when using the sorbent based on Ti$_{0.8}$Zr$_{0.2}$CrMn characterized by high isotope effects in comparison with LaNi$_5$, experiments in a column of length 0.26 m are carried out in the molar flow density range 4–15 mol/m^{2}s at 273 and 293 K. The results agree well with the data in Fig. 4.13. This indicates that even a short column (< 0.26 m) is sufficient to form a steady front for the IMC studied.

Figure 4.14 presents the dependences of h_{0g} on the specific flow for different pressures (from 0.15 to 1.9 MPa) and $T = 228$ K, whence it follows that a pressure increase results in an appreciable decrease of HTU for both sorbents studied.

To explain the experimental dependences on temperature and pressure, let us consider more comprehensively the components of h_{0g} [see (4.20)], which,

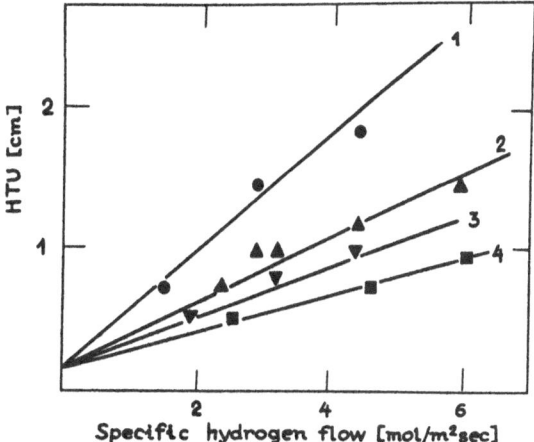

Fig. 4.12. Dependence of h_{0g} for the H–T mixture on the specific hydrogen flow for the sorbent on the basis of LaNi$_5$ at $P = 0.5$ MPa and different temperatures: (1) 228 K, (2) 250 K, (3) 273 K, (4) 293 K

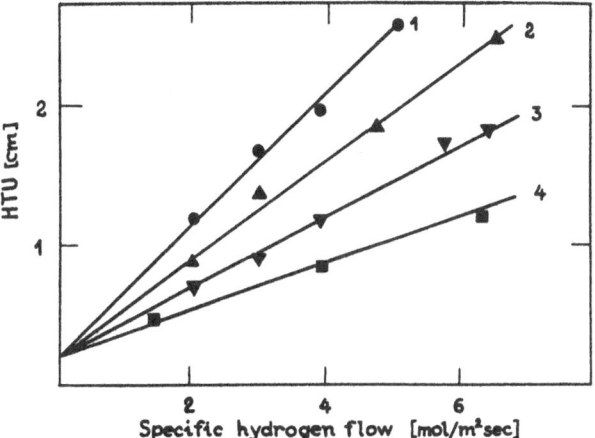

Fig. 4.13. Dependence of h_{0g} for the H–T mixture on the specific hydrogen flow for the sorbent on the basis of Ti$_{0.8}$Zr$_{0.2}$CrMn at $P = 1.9$ MPa and different temperatures: (1) 195 K, (2) 228 K, (3) 250 K, (4) 273 K

with respect to the above-considered steps (Sect. 4.6), can depend on diffusion of molecular hydrogen through the sorbent pores to the active surface of IMC. It was shown previously that the contribution of axial dispersion and external diffusion resistance can be neglected and, hence, the basic components of HTU are connected with hydrogen diffusion in the pores of the sorbent grains (h_p) and the transport of atomic hydrogen from the surface to interstices of the crystal lattice (h_m). Thus, (4.20) takes the form

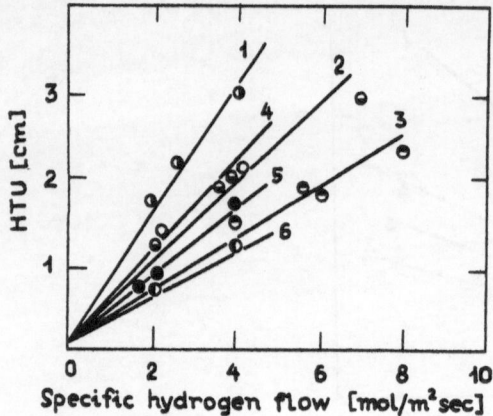

Fig. 4.14. Dependence of h_{0g} at 228 K for the H–T mixture on the specific hydrogen flow in the column filled with the sorbent on the basis of $Ti_{0.8}Zr_{0.2}CrMn$ (*1, 2, 3*) and LaNi$_5$ (*4, 5, 6*) at different pressures: 0.15 MPa (1, 4), 0.5 MPa (2, 5), 1.9 MPa (3, 6)

$$h_{0g} \approx h_s = h_p + \frac{\lambda h_m}{\alpha} = \frac{G_{sp}}{\beta_p a_{gr}} + \frac{\lambda L_{sp}}{\alpha \beta_m a_{gr}}, \tag{4.68}$$

where β_p and β_m are mass-exchange coefficients due to the processes of diffusion of molecular hydrogen in pores of the sorbent grains and transport of atomic hydrogen from the surface to interstices of the crystal lattice; a_{gr} is the mass-transfer surface determined by the geometrical surface of the particles (crystallites) of IMC; for sorbents based on LaNi$_5$ and $Ti_{0.8}Zr_{0.2}CrMn$ $a_{gr} = 5 \times 10^5 \, m^2/m^3$.

To estimate the contribution of each step it is necessary to use the data on the kinetics of isotope exchange on IMC powders shown in Table 4.3.

Using the value of activation energy to determine the temperature dependence of β_m and taking into account the temperature dependence of β_p, which is determined by the diffusion coefficient of molecular diffusion D_{H_2} (the mean path length of hydrogen molecules $\bar{\lambda}$ in the temperature range 228–293 K and in the pressure range 0.5–1.9 MPa varies from 4.9×10^{-9} to 23.0×10^{-9} m, i.e., $\bar{\lambda}/d_{pore} \ll 1$, which determines the range of molecular diffusion for which $D_{H_2} \propto T^{3/2}$), and by the experimental values of h_{0g} obtained for the H–T mixture at equal flow, and solving the set of equations (4.69), values of h_p and h_m were determined (and, hence, values of β_p and β_m:

$$h_{0g1} = h_{p1} + h_{m1} \quad \text{at } T_1 ,$$
$$h_{0g2} = h_{p2} + h_{m2} \quad \text{at } T_2 . \tag{4.69}$$

It is taken into account in these equations that values of HTU for the H–T mixture, determined by the above-described technique (Sect. 4.3) in the range of tritium trace amounts, are obtained at $\lambda = \alpha$. The calculations are carried out at $G_{sp} = 3 \, mol/m^2s$, i.e., in this case the dependence of HTU on the flow remains linear, indicating the absence of any influence of the hydrodynamic operating mode in the column on the coefficients β_p and β_m. The results of calculation are shown in

Table 4.8, whence it follows that at $T = 228\,K$ the rate of IIE is determined both by the step of molecular hydrogen diffusion in the pores of the sorbent grains and by the step of hydrogen atom implantation in the crystal lattice interstices. With increasing temperature, β_m increases faster than β_p and at $T = 228\,K$ for the sorbent based on LaNi$_5$, the component h_p makes the main contribution to h_{0g}. Since for LaNi$_5$ the activation energy is considerably higher than for Ti$_{0.8}$Zr$_{0.2}$CrMn, for the sorbent based on the latter IMC, the contribution of h_m to HTU remains essential even at moderate temperatures and pressures.

Table 4.8. Dependence of h_{0g} and β_{0g} and their components on temperature and pressure at the H–T exchange on granulated sorbents ($G_{sp} = 3\,mol/m^2 s$)

IMC	P MPa	T K	$h_{0g} \times 10^2$ m	α_{H-T}	$h_m \times 10^2$ m	$h_p \times 10^2$ m	$\beta_{0g} \times 10^5$ mol/m^{2}s	$\beta_m \times 10^5$ mol/m^{2}s	$\beta_p \times 10^5$ mol/m^{2}s
LaNi$_5$	0.5	228	1.43	1.65	0.53	0.90	42	69	67
		250	0.99	1.43	0.20	0.79	61	210	76
		273	0.78	1.27	0.10	0.68	77	470	88
		290	0.68	1.16	0.06	0.62	88	860	97
	1.9	228	1.26	1.65	0.39	0.87	48	93	69
		290	0.64	1.16	0.04	0.60	94	1290	100
Ti$_{0.8}$Zr$_{0.2}$ CrMn	1.9	195	1.93	3.30	0.86	1.07	31	21	56
		228	1.39	2.36	0.54	0.85	43	47	70
		250	1.07	2.02	0.33	0.74	56	90	81
		273	0.91	1.77	0.26	0.65	66	130	92
		293	0.81	1.59	0.22	0.59	74	170	100

Similar calculations have been carried out for the H–D isotope exchange, and the results are summarized in Table 4.9. It follows from comparison of the data in Table 4.8 and Table 4.9 that for the sorbent based on LaNi$_5$ efficiencies of IIE for H–T and H–D mixtures are similar and for the sorbent based on Ti$_{0.8}$Zr$_{0.2}$CrMn the isotope exchange H–T proceeds considerably more effectively than the H–D exchange ($\beta_m^{H-T} > \beta_m^{H-D}$). These results agree well with the data on temperature dependence of rates of isotope exchange R_{H-T} and R_{H-D} on IMC powders presented in Fig. 4.14.

It is significant that the coefficient β_p is practically independent of the type of isotope mixture (this is related to the similarity of the diffusion coefficients for the isotope modifications of molecular hydrogen), and of the IMC nature (since the character of the sorbent grain structure is the same), and of pressure. This means that, regardless of the listed factors, at constant temperature h_p is determined only by the flow in the column. For example, at $T = 228\,K$ and $G_{sp} = 3\,mol/m^2 s$, h_p appears to be equal to 0.85 cm.

Table 4.9. Dependence of h_{0g} and β_{0g} and their components on temperature at the H–D exchange on granulated sorbents ($G_{sp} = 3\,mol/m^2s$, $P = 1.9\,MPa$)

IMC	T K	$h_{0g} \times 10^2$ m	α_{H-D}	$h_m \times 10^2$ m	$h_p \times 10^2$ m	$\beta_{0g} \times 10^5$ mol/m²s	$\beta_m \times 10^5$ mol/m²s	$\beta_p \times 10^5$ mol/m²s
LaNi$_5$	228	1.39	1.36	0.57	0.82	43	77	73
	293	0.63	1.12	0.07	0.56	95	710	107
Ti$_{0.8}$Zr$_{0.2}$	195	3.29	2.40	2.24	1.05	18	11	57
CrMn	228	2.05	1.88	1.22	0.83	29	26	72
	250	1.64	1.67	0.92	0.72	37	39	83
	273	1.34	1.48	0.71	0.63	45	57	95
	293	1.16	1.36	0.59	0.57	52	75	105

Let us dwell on the dependence of β_m on pressure, which can be estimated using the experimental values of h_{0g} obtained at $T = 228\,K$ and shown in Fig. 4.14. Table 4.10 presents the values of β_m calculated with the information that, at this temperature and load, $G_{sp} = 3\,mol/m^2s$ and $h_p = 0.85\,cm$. For the sorbent based on LaNi$_5$ the following dependence is found: $\beta_m \propto P_{H_2}^{0.27}$; for the sorbent based on Ti$_{0.8}$Zr$_{0.2}$CrMn, $\beta_m \propto P_{H_2}^{0.33}$, which agrees satisfactorily with the dependence of the isotope exchange rate $\overline{R}$ for powders on pressure (Table 4.3).

Table 4.10. Dependence of h_m and β_m on temperature and pressure at the H–D exchange on granulated sorbents ($G_{sp} = 3\,mol/m^2s$)

P	LaNi$_5$			Ti$_{0.8}$Zr$_{0.2}$CrMn		
MPa	$h_{0g} \times 10^2$ m	$h_m \times 10^2$ m	$\beta_m \times 10^5$ mol/m²s	$h_{0g} \times 10^2$ m	$h_m \times 10^2$ m	$\beta_m \times 10^5$ mol/m²s
0.15	1.70	0.85	43	2.40	1.55	16
0.5	1.43	0.58	63	1.75	0.90	28
1.9	1.26	0.41	89	1.39	0.54	47

Thus, using both techniques to study isotope exchange it is possible to establish the rate-determining steps of the process and seek ways to enhance them.

5. The Use of H$_2$–Me(IMC) Systems for Hydrogen Isotope Separation

5.1 Periodic Separation Processes

Despite their considerable separation effects, the H$_2$–Me(IMC) systems are not widely used because of the difficulties in realizing continuous counter-current separation. Thus, most work on hydrogen isotope separation in columns is devoted to periodic separation processes and primarily to chromatographic separation. Chromatography can provide a greater separation degree and, hence, it is very interesting for solving analytical tasks related to monitoring the tritium content in the environment or determining the deuterium concentration at natural levels and below.

When a mixture of hydrogen isotopes passes through a column filled with palladium, enrichment of the forward front with the heavy isotope occurs. There are two alternative schemes for the process: In the first, the gas flow in the column is produced by a continuous feed of hydrogen; in the other, it is produced by hydrogen displacement from the palladium bed by heating the column element after preliminary saturation by hydrogen.

Chromatographic separation of hydrogen isotopes in the H$_2$–Me system was first performed by *Glueckauf* and *Kitt* [5.1, 2]. They used a column of length 0.44 m and diameter 0.8 cm filled with Pd powder (20 g) mixed with asbestos (6 g). The column operated on the principle of thermal desorption and made possible the extraction of deuterium of purity 99.15 % with a yield of > 95 % from the mixture of hydrogen isotopes containing 40 at% of deuterium. In addition to the determination of α the chromatographic separation was used for analytical purposes, namely, enrichment of hydrogen obtained from water samples with tritium. *Hoy* [5.3] devised a method in which a water sample of mass $\approx$ 50 g was reduced to hydrogen by magnesium at a temperature of 923 K. About 70 l of hydrogen was fed into the previously evacuated chromatographic column filled with palladium powder of mass $\approx$ 1 kg. Transport of sorbed hydrogen along the column is achieved by heating up to 520 K. The fraction enriched with tritium was fed into a second column also filled with Pd (359 g). Thereafter the enriched fraction entered a third (33 g of Pd) and fourth (4 g of Pd) column. Finally, it has been possible to concentrate $\approx$ 60 % of the tritium contained in the source sample into the enriched hydrogen fraction of volume 0.5 l (this corresponds to $\approx$ 0.7 % of the initial sample volume).

A relatively low separation degree (≈ 90) is the drawback of this method, since a complicated device with a large quantity of Palladium ($\approx 1.4\,\text{kg}$) is required. *Tistchenko* and *Dirian* [5.4] applied the method of frontal chromatography for enriching hydrogen samples with tritium. Pd ($\approx 20\,\text{mass\%}$) on α–Al_2O_3 is used as the sorbent. The column is initially filled with inert gas (helium) which is blown out by hydrogen during the sorption. The authors obtained a separation degree of 100 in a column of length 1.12 m and diameter of 10 mm for an enriched fraction volume of $5\,\text{cm}^3$. They showed that the dependence of the separation degree on the withdrawn volume has exponential character.

The experiments reported in [5.5] show that the sorbent Pd–Al_2O_3 is less effective in isotope exchange than granulated palladium powder (at $T = 294\,\text{K}$, $P = 0.1\,\text{MPa}$ and a specific hydrogen flow $G_{sp} = 3 \times 10^{-2}\,\text{mol/m}^2\text{s}$, values of HTU for the sorbent containing 5 mass% of Pd on Al_2O_3 are found to be 6 times higher than for granulated palladium powder).

Another drawback of the technique considered is related to the fact that hydrogen is fed into a column that is initially filled with inert gas to avoid a pressure gradient in the column and to avoid the related effect of isotope axial dispersion. Injection of helium into the column prevents a significant pressure gradient and, hence, results in an increase of separation degree. However, it considerably complicates the procedure of enrichment and subsequent gas analysis, since an additional analytical device is required to note the moment when the helium content at the output of the column falls to zero (a device for heat conduction measurement is used in [5.4]). Furthermore, the enriched gas sample is contaminated by helium, which constrains the possibilities of subsequent analysis and requires other methods independent of the helium presence in the sample (mass-spectrometry is used in [5.4]). If samples without the inert gas are required, the loss of the richest part of the samples is inevitable because extraction of hydrogen must be delayed until after the helium output from the column is completed. In view of the abrupt exponential form of the concentration profile of the desired component in the enriched flow, it is necessary to rigorously note the moment of sampling in order to achieve maximum reproducibility of the separation degree. This also hampers the use of the technique described.

If the separation process is carried out at lower temperatures, the above drawbacks can be eliminated. In this case one attains not only a decrease of the pressure gradient in the region of the sorption front due to a decrease of the α–β transition pressure, but also an increase of the separation factor and of the hydrogen capacity of Pd. The last fact enables one to increase the sample volume or to decrease the size of the separation column and thus the amount of palladium in it, all other factors being the same. However, a decrease in the rate of interphase isotope exchange of hydrogen is also observed, along with the above-mentioned advantages, at reduced temperatures. This leads to an extreme character of the temperature dependence of the separation degree in the column.

These considerations about the influence of temperature on the degree of hydrogen isotope separation in a chromatographic column were experimentally checked in [5.6]. In addition, the influence of the enriched fraction volume and the gas flow on the separation degree was also investigated.

Experiments were carried out using the apparatus presented schematically in Fig. 5.1. The isotope mixture is fed continuously from the cylinder (1) through the device (2) for pressure check and traps (3,4) with liquid nitrogen into the chromatographic column (6) of diameter 7 mm, previously evacuated at 500 K to a residual pressure of about 6.5 Pa. The column is filled with Pd powder. Hydrogen flow is measured by means of the device (5). The temperature in the column is maintained using a thermostating liquid in the vessel (7). The first fractions of hydrogen enriched with heavy isotope are collected in the calibrated volume (10), from which the sample is taken in the evacuated container (8). The pressure in the container is checked by means of a U-shaped manometer (9). Analysis of deuterium in enriched samples of hydrogen is performed by spectral methods. A chromatographic column of length 0.6 m filled with granulated palladium powder (mass of Pd 136 g, grain size 0.2–0.5 mm) is used in the experiments.

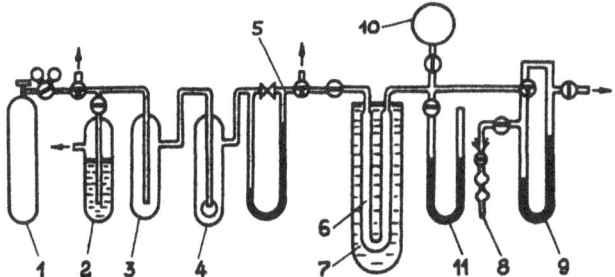

Fig. 5.1. Scheme of laboratory chromatographic plant with palladium sorbent [5.6]: (1) isotope mixture storage vessel, (2) outlet device, (3,4) lidquid nitrogen cold traps, (5) flowmeter, (6) chromatographic column, (7) thermostated bath, (8) outlet line to an evacuated container, (9) manometer, (10) vessel with calibrated volume

The experimental results are presented in Table 5.1. Analysis of the results shows the possibility of a considerable increase of the separation degree by decreasing the temperature in the chromatographic column to 273 K and the possibility of a reliable reproducibility of the enrichment results (e. g., experiments 2, 3, and 5, 6 in Table 5.1).

In summery it should be noted that precise knowledge of the separation degree attained is required for the determination of the tritium concentration in the source hydrogen using the result of enriched samples analysis. In this connection preliminary calibration of the column is required. However, it is possible to avoid this by noting that the deuterium concentration in natural objects is constant (0.015 at%) and by performing, together with the analysis for tritium, an analysis of deuterium in the enriched sample.

To conclude this section we note that, despite the drawbacks of the periodic separation process, a pilot chromatographic plant for hydrogen isotope separation

Table 5.1. Results of experiments on chromatographic separation of protium–deuterium mixtures on palladium

Experiment	T	G	Enriched fraction volume	Deuterium concentration (at %)		Separation degree
				in source gas	in enriched sample	
number	K	mol/m^{2}s	cm^3			
1	299	0.88	0.62	2.5	49	37.5
2	302	1.10	0.50	2.5	52.5	43.1
3	303	1.10	0.62	2.5	54.0	45.7
4	273	1.22	1.80	2.5	> 98	–
5	273	1.13	1.74	0.015	13.0	997
6	273	1.22	1.86	0.015	13.6	1040
7	253	1.24	0.62	0.015	1.8	120

based on the isotope effect in the H$_2$–Pd system has been constructed in France [5.7], and recently the possibility of applying the gas chromatography method for separation of isotopes from a thermonuclear reactor has been considered. The system for hydrogen isotope separation was designed in the framework of the JET program, and is intended for the extraction of 5 moles of T$_2$ and 15 moles of D$_2$ per day. It consists of four chromatographic columns of volume 5 l filled with palladium on Al$_2$O$_3$ [5.8, 9].

The Efficient Palladium Isotope Chromatograph for Hydrogen (EPIC) constructed in Mound Applied Technologies with the use of Pd on α–Al$_2$O$_3$ and experiments using it for separation of a H$_2$–D$_2$ mixture are described in [5.10]. The chromatograph is equipped with a computer to minimize the cycle duration and obtain isotope purities > 99.9 at%. An estimation of the separation efficiency of a four-component mixture (H, D, T, and ^{3}He) is also reported in [5.10].

The method with pressure variation (Pressure Swing Absorption Process – PSAP) and the method with thermal cycling (Thermal Cycling Absorption Process – TSAP) are classified as semicontinuous methods. Both represent semicontinuous chromatographic processes and have been used for the separation of hydrogen isotopes.

Waever and *Hamrin* [5.11] applied the four-stage scheme with adsorption without heating (suggested in [5.12]) for enrichment of a H–D mixture containing 5.50 at% of D. Therewith Pd pressed with Al (25 mass% of Pd) was used as a sorbent. Two columns of length 87.6 cm and diameter 0.94 cm were used in the experiments. Each column contained 95 g of the sorbent. The experiments were carried out at room temperature (298 K) and constant hydrogen pressures: high – 0.790 MPa, and low – 0.136 MPa, corresponding to the β-phase of Pd. Let us explain the essence of the method using the scheme presented in [5.11-13] and shown in Fig. 5.2.

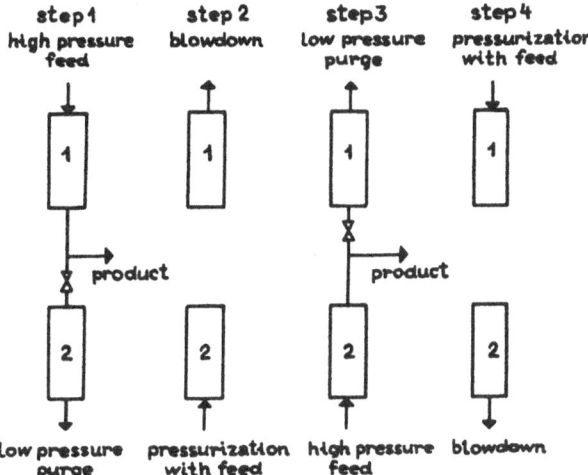

Fig. 5.2. Scheme of stages of the adsorption process with varying pressure [5.13]

In the initial state column 1 is kept under high pressure (P^H) and column 2 at low pressure (P^L). In both columns absorption and isotope equilibrium between the gas phase and palladium hydride is established. In the first stage the feed flow passes through column 1 and is enriched with the heavy isotope during isotope exchange (in the case of Pd sorbent). At the output of column 1 part of the flow is withdrawn as a product and the remainder is reduced and enters column 2, where it is also enriched with the heavy isotope. Thereafter the high pressure feed gas enters column 2, where it is additionally sorbed. At the next stage column 2 fed with the source gas, where it is enriched with deuterium, which is withdrawn as a product. The remainder enters column 1, in which the pressure is reduced. Passing through column 1 the flow is depleted of the light isotope. The method is presented in detail in [5.11, 13]. In [5.13] the hydrogen–V hydride system is studied using a protium–tritium mixture.

The method is based on the change of conditions of sorption and isotope equilibrium, since the separation factor is practically independent of pressure, and the capacity for hydrogen depends only slightly on pressure in the range of the β-phase. The efficiency of this method is very low. Table 5.2 presents the data on the source mixture enrichment obtained in [5.11, 13] for various experimental conditions.

In the general case it may be concluded that the maximum enrichment is achieved for minimum cycling time $\Delta\tau \to 0$, maximum difference in pressures, and minimum product withdrawal. However, considerable mixing effects reduce the attainable enrichment.

The method of thermal cycling, applied for hydrogen isotope separation in [5.14, 15], is more effective. The scheme of the column disposition in the absorption process with thermal cycling is shown in Fig. 5.3.

Table 5.2. Experimental data on hydrogen isotope separation by the PSAP method

Sorbent	y_B^*/y_0	$\Delta\tau$	T	The number of semicycles	$P_{H_2}^H/P_{H_2}^L$
		s	K	n	
β-Pd on Al	1.112	30	298	20	5.82
	1.054	60		20	
	1.045	120		20	
$VH_{0.7}$	1.75	18	373	300	10
	1.64	30		200	
	1.62	60		100	
	1.47	120		50	
	1.33	240		25	
	1.28	360		16	

y_B^* is the heavy isotope concentration in the product

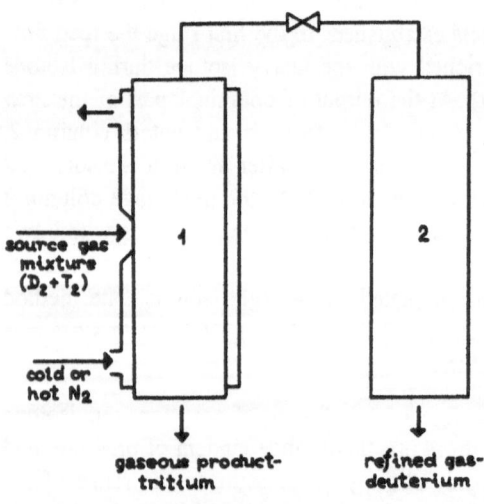

Fig. 5.3. Scheme of columns disposition in the absorption process with thermal cycling [5.14]

Column 1 is filled with palladium on kieselgur; this column can be heated and cooled. Column 2 is empty. The upper part of column 1 is connected by a bolt (or valve) to column 2, which is the column with alternating direction of the gas flow.

Refined gas consisting of light hydrogen isotopes is withdrawn from the upper part of column 1 and gas enriched with tritium from the lower part.

The process cycle involves half-cycles of cooling and of heating. Figure 5.4 shows the basic stages of the process. During the cooling half-cycle, because of hydrogen isotope absorption by palladium, the mixture of isotopes passes from

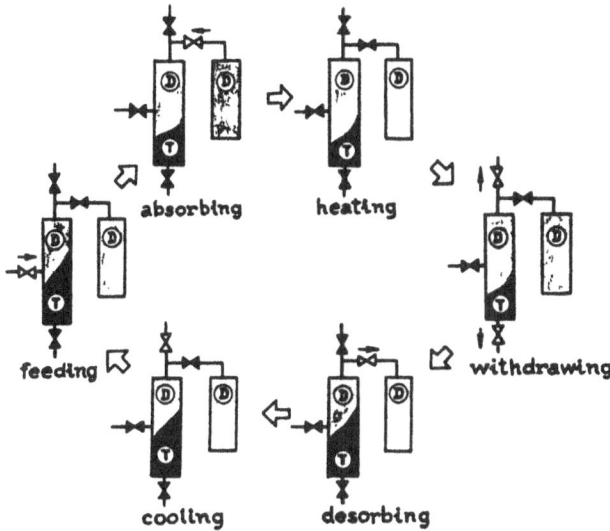

Fig. 5.4. Basic stages of the process with thermal cycling [5.15]

column 2 to column 1; therewith enrichment of the gas with the heavy isotope, tritium, occurs. Tritium is concentrated in the lower part of column 1, whence it is withdrawn as a product. During the heating half-cycle the gas previously sorbed by Pd is desorbed from the solid phase and passes from the lower part of column 1 to column 2. In so doing the gas flow is also enriched with the heavy component. However, since at high temperatures α is considerably lower, this only partially reduces the separation effect attained during the cooling half-cycle. So an overall gain in separation is achieved during the full cycle. A product of high concentration is obtained by repeated cycling.

To maintain the semicontinuous operating mode, at the beginning of each heating half-cycle minor portions of the gas from column 1 are withdrawn from its upper and lower parts and at the beginning of each cooling half-cycle an equal amount of the source gas mixture enters the middle part of column 1 to keep a constant amount of gas. The productivity of such a process depends on the required purity of the endproduct. The work [5.15] contains data on enrichment of the H–D mixture (55 % protium and 45 % deuterium) up to 9.5 %. The presence of considerable mixing effects and the semicontinuous character of the separation are the common drawbacks of this and the previously described process. In conclusion it is to be noted that when using these methods, it is necessary to coordinate the time of the half-cycles with the rate at which phase and isotope equilibrium are established in hydrogen-isotopes–metal-hydride/IMC systems.

5.2 Continuous Counter-Current Separation Processes

Continuous counter-current separation processes are widely used for separation of hydrogen isotopes and other light elements so far as gas (vapour)–liquid systems rectification processes, chemical isotope exchange, and two-temperature methods are considered. In this case, as a rule, schemes with current reflux and two-temperature schemes (GS-process) are widely used in industrial hydrogen isotopes separation plants for heavy water production. Realization of such processes is in principle possible for systems with a solid phase.

First let us consider variants of a simpler scheme with flow conversion shown in Fig. 5.5. The main parts of the plant are: separation column (1), desorber (2), and absorber (3) for the conversion of gas and solid phase flows. In systems with a positive isotope effect, the heavy isotope (e. g., tritium) is concentrated in the lower part of the column. Hence, the enriched product is withdraw from the gas emerging from the desorber (2). Waste hydrogen flow is withdrawn from the gas flow before it enters the absorber. If the desired isotope is concentrated in the gas phase (e. g., tritium in systems with a negative isotope effect), the product is withdrawn from the upper part of the plant and the hydrogen component released in the desorber forms the waste. The separation column has concentration and depletion sections (for the desired isotope) and hence the source flow is fed into the middle part (Fig. 5.5a).

In the scheme considered a flow conversion occurs at the upper end of the column (in the absorber) through hydrogen sorption by the solid phase and at the lower end (in the desorber) by way of hydrogen desorption from the solid (hydride) phase as a result of heating (or e. g., blowing by inert gas or decrease of pressure over the solid phase). Coolant is fed to the absorber to remove the heat of hydride formation (ΔH_H). The lower flow conversion, as a rule, results from

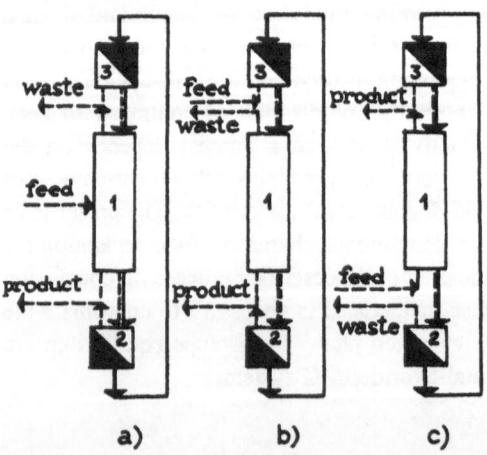

Fig. 5.5a–c. Scheme of separation plants using hydrogen – solid phase systems: (a) scheme with concentration and depletion, (b, c) open scheme for heavy isotope concentration in systems with positive and negative isotope effect, respectively

thermal desorption. Solid phase heating in the desorber is achieved by means of a heat-transport medium. Regenerated solid phase in the desorber is fed to the absorber.

The flow conversion can be executed either directly in the upper and lower parts of the counter-current column (i.e., the column consists of three zones: separation, sorption and desorption) or in separate apparatuses.

If the plant works according to the open scheme, only one flow-conversion unit is demanded: the lower unit when using a solid phase in which the desired isotope is concentrated (Fig. 5.5b), or the upper one for concentration of the heavy isotope in a gas phase (Fig. 5.5c). However, in the first case an absorber is also required, since the column is fed by solid phase containing hydride; and in the second case a desorber is necessary for producing the waste hydrogen flow and regeneration of the solid phase, which subsequently enters the upper flow-conversion unit. Thus separation plants using the H_2–Me(IMC) systems are characterized by the fact that both absorber and desorber are essential when operating under an open scheme (though only one of the two is a flow-conversion unit).

5.2.1 Hypersorption

The main difficulties in the practical application of solid–phase systems for isotope separation concern the realization of counter-current movement of gas and solid phase. The first variant provides for separation in a column with a dense bed of solid phase which falls under gravity. Hydrogen passes through the bed in the opposite direction. In the flow-conversion units counter-current phase movement can remain. Such counter-current separation processes are often referred to as hypersorption.

The hypersorption process was originally used for the separation of the isotope mixture H_2–D_2, in which hydrogen was sorbed by activated carbon [5.16] and silica gel [5.17]. Thereafter it was realized in the H_2–Pd system for separating the H_2–HT mixture [5.18]. The column filled with granulated palladium sorbent (spherical grains of diameter 1.3 mm) is the main part of the apparatus illustrated in Fig. 5.6 [5.18, 19]. The velocity of the solid phase (palladium hydride) falling under gravity is controlled by means of a disk machine. The solid phase removed from the disk machine by a special device falls in the lower container, which acts as desorber. Hydrogen desorption is performed at a temperature of 520–530 K. Once the container has been filled with palladium sorbent, it is evacuated and hold at the upper end of the column.

The plant can operate both with the upper unit for gas flow conversion (regime of heavy hydrogen isotope concentration) and with the lower unit for hydride phase flow conversion (regime of light hydrogen isotope concentration). In the first case, the source mixture of hydrogen isotopes, after removal of impurities, is fed into the disk machine, passed through the column from the bottom upwards and sorbed in the upper part by regenerated palladium entering from the upper container. Hydrogen desorbed from palladium in the lower container is removed from the plant. In the second case, palladium (in the form of hydride) from the

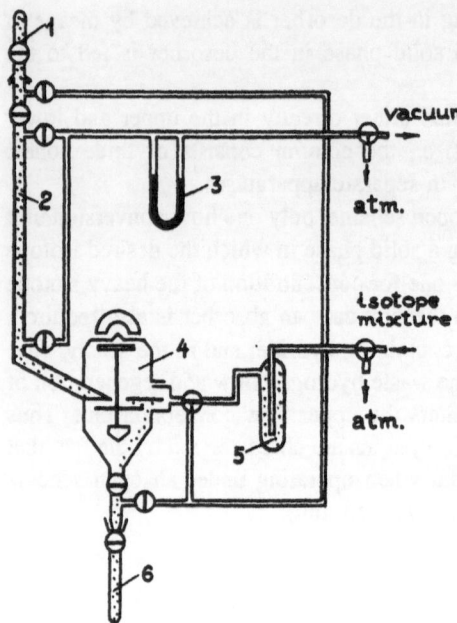

Fig. 5.6. Scheme of the experimental hypersorption plant with granulated palladium sorbent [5.18, 19]: (1) absorber, (2) separation column, (3) flowmeter, (4) disk machine, (5) purification system, (6) desorber

upper container saturated by the source mixture is fed into the column and all the desorbed hydrogen is returned to the column. The extent of desorption is checked by the waste hydrogen flow leaving the plant.

Table 5.3. Characteristics of experimental hypersorption plants for separation of hydrogen isotopes

Sorbent	Mixture separated	Temperature	Loading G_{sp}	Column sizes		Separation degree	HETP
		K	kmol/m^{2}h	height	diameter		cm
Palladium	H$_2$–HD	294	1.8	20	1.5	> 122*	< 2.5
Silicagel	H$_2$–D$_2$	77	3 m/h**	200	2.0	56***	6.75***
Activated carbon	H$_2$–D$_2$	86	2.5–12.6	46	3.8	42	1.6

* value obtained until the permanent state in the column
** linear velocity of the sorbent in the column
*** values relate to the concentrating part, HETP in the depleting part is equal to 4.75 cm

The column sizes, the working parameters, and the separation efficiency are summarized in Table 5.3, in which the characteristics of hypersorption separation plants employing silicagel and activated carbon as hydrogen sorbent are also presented. As is evident from Table 5.3, the highest separation efficiency is observed for the plant with palladium.

5.2.2 Sectioned Column

The hypersorption process is typified by a number of drawbacks related to the movement of a solid phase: sorbent attrition, decrease in the separation efficiency due to axial dispersion in the solid phase and difficulties related to sorbent dosage and transport. That is why the continuous counter-current separation process in column with an immobile bed of solid phase was proposed [5.20] and experimentally realized in several variants. The proposed method involves counter-current movement of the phases which can be realized not only by their physical transport but by the displacement of the flow-conversion units (or temperature zone in the case of thermal desorption) with reference to the immobile solid phase. The process is shown schematically in Fig. 5.7 which pictures a column consisting of several equal sections. Figure 5.7a presents an instant at which section 1 is the desorber and section 5 the absorber. Gas flow through sections 2–4 is achieved by elevated hydrogen pressure in the desorber and hydrogen sorption in the absorber. In passing through the separation sections the gas exchanges with hydrogen in the solid phase and is subsequently sorbed in section 5. Assume that the section is filled with palladium sorbent; in this case, hydrogen is enriched with heavy isotope upon isotope exchange with the hydride phase. Once the sorption has been completed, flow-conversion units displace the sections along the direction of gas movement, i.e., section 2 becomes desorber, section 1 becomes absorber, and sections 3–5 become separation zone. As a result, counter-current movement of solid phase L occurs with reference to gas flow G. Figure 5.7a shows the basic scheme of the column operating without product withdrawal. Figures 5.7b,c present the variants of the open scheme similar to those considered above and shown in Figs. 5.5b,c. When using a Pd sorbent, the scheme in Fig. 5.5b maintains the column in operation according to the depletion regime (purification of heavy isotope, for example,

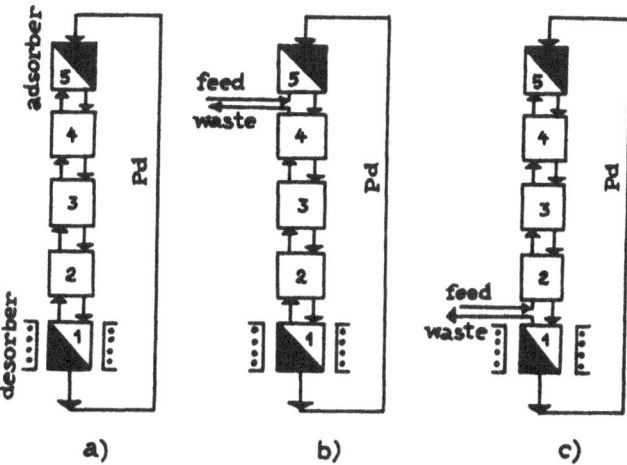

Fig. 5.7a–c. Principle of phase counter-current realization in columns with the solid phase immobile with respect to the apparatus walls

tritium), and the scheme in Fig. 5.5c corresponds to the regime of heavy isotope concentration. During plant operation with feed and waste, the points of feed and withdrawal of product and waste must be displaced synchronously, along with the above-considered movement of the flow-conversion units for the regime with no product withdrawal.

In the second variant, sections separated from another move in the direction opposite to the gas flow. That is, once the gas sorption has been completed, section 2 becomes desorber and section 1 becomes absorber as a result of the displacement.

It is evident that in both variants of counter-current movement realization, in addition to the schemes considered in Fig. 5.7, it is possible to carry out separation in a column with depletion and concentration sections according to the scheme shown in Fig. 5.5a.

The suggested process of counter-current separation was originally realized with real displacement of sections with respect to the flow-conversion units. This is shown schematically in Fig. 5.8 [5.21]. The separation column is the major part of the plant. It operates according to the open scheme without withdrawal of product and consists of seven sections (1–7) separated from one another by vacuum valves. Each section is connected to the adjacent one by means of pins. All sections of diameter 10 mm are filled with granulated Pd powder (grain size 0.3–0.5 mm) to the bed height of 2 cm. Compared to a hypersorption plant the requirements on sorbent mechanical strength are less rigid, enabling one to use just palladium powder. The technique was the following: after the six forward sections have been saturated by source isotope mixture H–D, the mixture is fed into the column through the lower section 1. Gas feed during operation and saturation of palladium by hydrogen are performed through traps 8 and 9 cooled by liquid nitrogen (for the purpose of gas purification and water removal), through device (10) for maintenance of gas flow constancy, and through the flowmeter 11.

Once palladium has been saturated by hydrogen, valves a, b and c are closed in section 7, section 1 is disconnected from the remainder, and the gas feed system is connected with section 2 by means of pin 13. Thereafter gas feed is restored. Section 1 is removed from the top of the column, placed in the kiln for desorption at 473 K, and evacuated. The source gas is enriched with deuterium when passing through the column from the bottom upwards. Once the upper cold section has been saturated, the displacement is repeated.

After attainment of steady state in the column (no more than 8 hours) the sections are disconnected, hydrogen is desorbed from palladium hydride and subsequently analyzed for isotope composition. It is thus possible to obtain both isotope concentrations in the solid phase (averaged for separate section) as a function of the column height and steady separation degree. The experimental results are summarized in Table 5.4.

Similar operation occurs in the plant in which counter-current movement is realized by displacement of sorption zones in the direction of gas flow with respect to immobile sections with Pd sorbent. Experiments on separation of the H–D mixture without withdrawal of product were carried out in this plant (Fig. 5.7a). When the column operated with circuited flows (without external feed), the desorber was

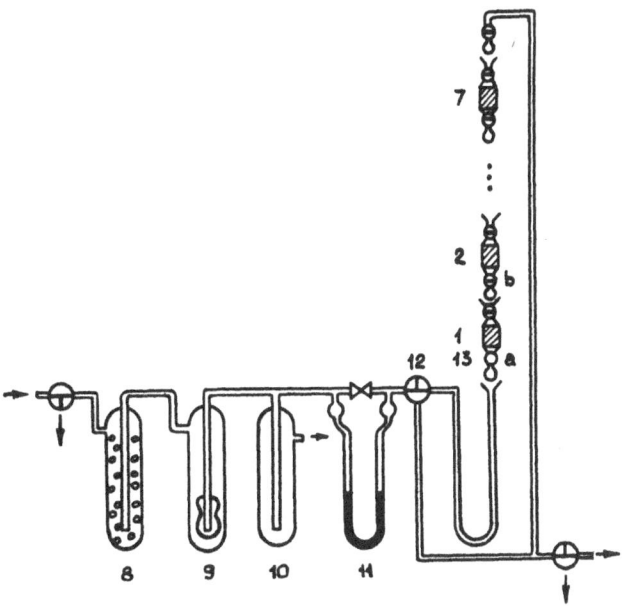

Fig. 5.8. Scheme of experimental separation plant with sectioned column filled with granulated palladium powder [5.21]: (1, 2–7) seven mobile sections, each filled with granulated Pd powder, (8, 9) liquid nitrogen cold traps, (10) buffer volume, (11) flowmeter. (See text on page 124)

Table 5.4. Results of the H–D mixture separation in sectioned column filled with palladium sorbent operating according to the open scheme ($T = 298$ K, $P = 0.1$ MPa [5.21])

Loading G_{sp}	Height of separation part of column	Deuterium concentration (at %)		Separation degree	HETP
		in source mixture	in product		
kmol/m^{2}h	cm			K	cm
1.1	8	8.5	80.0	43	2.1
2.2	8	8.5	79.5	42	2.1
2.4	6	20.0	80.0	16	2.1
2.9	6	20.0	68.0	8.5	2.7
3.4	6	8.5	67.0	22	2.1

not evacuated, thus avoiding gas loss. That is why one finds a dependence of the separation degree on operating pressure, which is related to the incompleteness of flow conversion in the desorber (the fraction of hydrogen desorbed in the free volume of the section, including solid bed, is placed in the sorption zone together with the section). In the column with palladium, protium is concentrated in the desorber, resulting not only in some change of the ratio of flows of gaseous hydrogen and dissolved hydrogen in palladium but also in isotope dilution at the column end enriched with deuterium (in the absorber). These drawbacks result from the above-mentioned gas bypass. It is evident that the incompleteness of flow conversion increases with increasing pressure and leads to the decrease in separation degree.

A subsequent improvement of the plant involved a decrease of the free space in it and automation of the process. This plant [5.22, 23] is equipped with an automatic system for the displacement of the sections and automatic control and maintenance of hydrogen pressure and flows.

For this purpose sections (in the form of cylinders of diameter 11 mm) are connected by a common device for flow division and distribution, which is leak-proof at pressures from 0.01 to 0.5 MPa. The rotable part of the device is connected to the sections filled with granulated Pd sorbent (grain size ≈ 1 mm); the immobile part includes external pipes, which are joined to the sections through additional connections.

Separation was carried out at atmospheric pressure in the temperature range 273–333 K. Hydrogen desorption from palladium occurred at a temperature of 473 K. The column consists of five sections, three of them being in the separation zone. A number of experiments were carried out in the column equipped with five sections. We studied separation of the isotope mixtures H–D, H–T, and D–T. Once steady state in the column is reached, the separation degree $K = x_R(1 - x_P)/([x_P(1 - x_R)]$ was determined from the isotope composition of the gas at "rich" (x_R) and "poor" (x_P) ends of the column. The concentration profile in the column was also plotted from the average isotope concentrations in the solid phase. The deuterium concentration, as in other sectioned plants, was found by spectroscopic methods from atomic hydrogen spectra and the tritium concentration was determined radiometrically by means of a counter with internal filling.

An essential feature of the considered version of counter-current separation is related to the periodic movement (real or relative) of the solid phase. This impairs the separation efficiency, i.e., leads to an increase of HETP as calculated from the experimental data with increasing height of the solid bed. This dependence is studied for the example of H–D separation (Table 5.5). As is obvious from the table, an increase in the height of the palladium sorbent bed in a section by a factor of 10 results in a HETP rise by nearly a factor of 7. The overall separation efficiency in the column appears to be high: for the sorbent bed of height 3 cm, $K = 500$, we failed to determine the separation degree in the column with 5 sections, since the deuterium concentration at the "poor" end of the column is found to be lower than the limiting response of the applied technique of isotope analysis. Hence, the majority of the experiments are carried out in a plant with three sections.

Another interesting result is the abrupt impairment of the separation efficiency for the H–D mixture on Pd sorbent discovered for deuterium concentrations above 98 %. As is obvious from Fig. 5.9, the dependence of HETP on the height of the palladium sorbent bed in a section remains linear.

Table 5.5. Experimental data on the dependence of the efficiency of the H–D mixture separation on the height of the palladium sorbent layer in a section at $T = 296$ K [5.22]

Height of sorbent layer cm		Deuterium concentration (at %)		Separation degree	HETP
in section	in column	x_P	x_B		cm
1.0	3.0	5.7	96.9	500	0.4
2.0	6.0	2.4	93.3	1566	0.5
5.6	17.0	2.4	97.7	1730	1.2
10.6	37.0	> 0.2*	93.5	> 15 600	2.7

* limiting response of the applied technique of isotope analysis

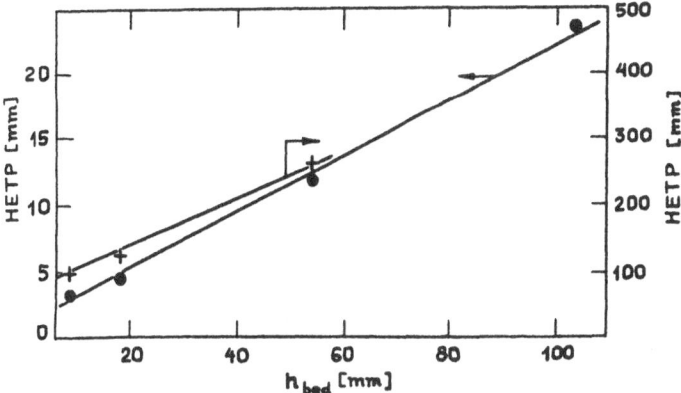

Fig. 5.9. Dependence of HETP on the height of the palladium sorbent bed in section at $T = 296$ K, $G_{sp} = 1.3$ mol/m²sec, and deuterium concentrations of: • less than 98 at%, + higher than 98 at% [5.22, 23]

Figures 5.10, 11 demonstrate that, in the studied range of change of separation conditions, HETP does not depend on the specific flow and temperature in the column for deuterium concentrations less than 98 at%. In the range of higher deuterium concentration the character of HETP dependence on the specific flow and temperature shows that the separation efficiency is determined by the rate of chemical reaction of hydrogen isotope exchange on the surface of the solid phase.

For comparison experiments on H_2–D_2 separation have been carried out in this plant using zeolite NaX as a hydrogen sorbent [5.23]. It is obvious from the data presented in Fig. 5.10 that both on palladium (for deuterium concentrations < 98 %) and on zeolite the efficiency of mass-transfer is equal, i.e., it does not depend on the nature of the sorbent.

The results of these experiments on the separation of H–T and D–T mixtures do not exhibit any surprising features. The values of HETP appears to be 3–4 mm, as in the case of separation of the H–D isotope mixture for a Pd sorbent bed in

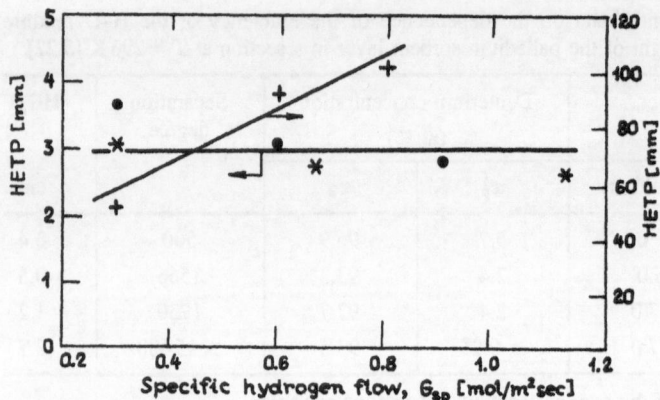

Fig. 5.10. Dependence of HETP on hydrogen flow at T = 296 K, H_{bed} = 10 mm, and deuterium concentration of: • less than 98 at%, + higher than 98 at%; * and for zeolite NaX at T = 78 K [5.23]

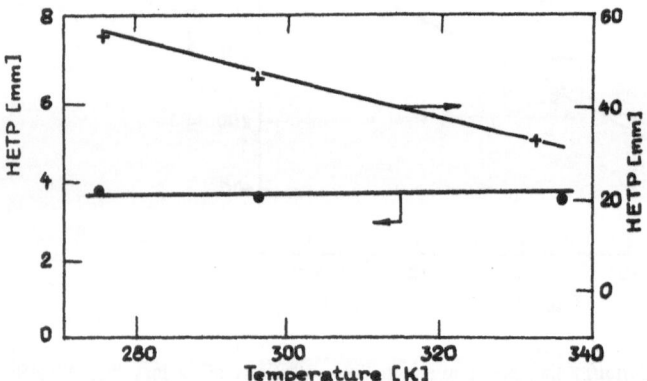

Fig. 5.11. Dependence of HETP on temperature at H_{bed} = 10 mm and deuterium concentrations of: • less than 98 at% (G_{sp} = 0.4 mol/m^{2}sec), + higher than 98 at% (G_{sp} = 0.3 mol/m^{2}sec)

a section of height 10 mm. The characteristic profile of the tritium concentration (specific activity) in the solid phase of the column is presented in Fig. 5.12. Such a profile is typical of counter-current columns.

In the case of separation of isotope mixtures with high tritium content, the gas can contain large amounts of radiation-generated helium, which are together with other impurities not absorbed by the solid phase, helium is concentrated in the sorption section. The influence of helium on the efficiency of the plant for tritium concentration has been experimentally explored. Figure 5.13 presents the dependences of gas pressure in the sorption section on time obtained for the H–T mixture containing helium at 5×10^{-2} at%. It is evident that at the beginning of the operation, the pressure in the sorption section after the displacement of

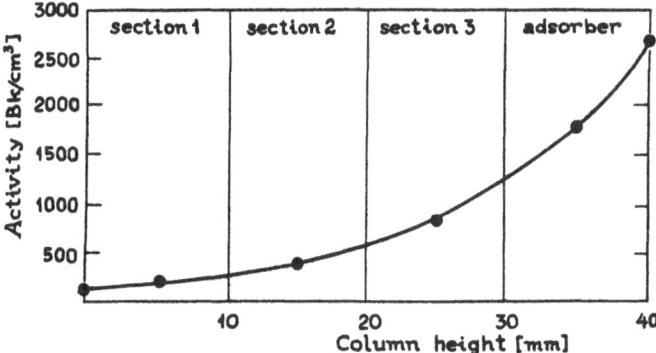

Fig. 5.12. Steady concentration profile of tritium in the hydride phase of palladium when separating the D–T mixture in sectioned column (T = 293 K, H_{bed} = 10 mm, G_{sp} = 0.6 mol/m^{2}sec)

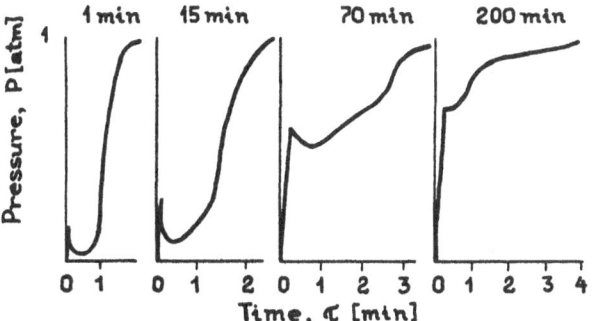

Fig. 5.13. Influence of helium impurity on the absorber efficiency in a plant for hydrogen isotope separation with sectioned column filled with palladium sorbent [5.22]

sections corresponds to the pressure of hydride formation. As helium is collected, the pressure of sorption increases and sorption becomes prolonged. Within 70 min of the start, the hydride formation time increased by nearly a factor 4, and after 5 h hydrogen sorption is interrupted due to the "gas cushion" effect. Periodic removal (evacuation) of inert gas from the sorption zone is found to be sufficient to correct this adverse effect.

5.2.3 Two-Temperature Method

The two-temperature method is a further possibility to realize continuous counter-current separation in systems with a solid phase. The two-temperature method based on the temperature dependence of the separation factor enables one to realize separation without using flow-conversion units, which usually determine the basic consumption of energy and materials. The idea behind this method can be demonstrated by the example of a simple scheme of heavy hydrogen isotope concentration shown in Fig. 5.14. The upper part of column 1 is fed by flow G of the substance in which the heavy isotope is concentrated (in conditions of thermodynamic equilibrium). The lower part of column 2 is fed by flow L of another substance moving in the opposite direction. The counter-current separation process occurs in column 1 at temperature T_1 with the separation factor α_1 as well as in columns with routine flow conversion. Since the heavy isotope is concentrated in flow G, its concentration is maximum in the lower part of column 1.

In the routine separation process, the flow G from the lower part should be fed to the flow-conversion system and in the two-temperature method this flow, enriched with the desired isotope, enters the second column operating at temperature T_2 ($\alpha_2^{-1} < \alpha_1^{-1}$). Thus, the transfer of heavy isotope from flow G into flow L occurs in column 2. This isotope returns to column 1 together with flow L. Thus flow-conversion units are not required, since column 2 operating at T_2 serves their functions. The efficiency of the separation plant is determined by the separation factors α_1, α_2, flow ratio $\lambda = L/G$, and the column height [5.19, 24, 25]. Figure 5.14 also presents the x–y diagram of the separation process executed in the range of low content to heavy isotope ($\alpha \approx x/y$).

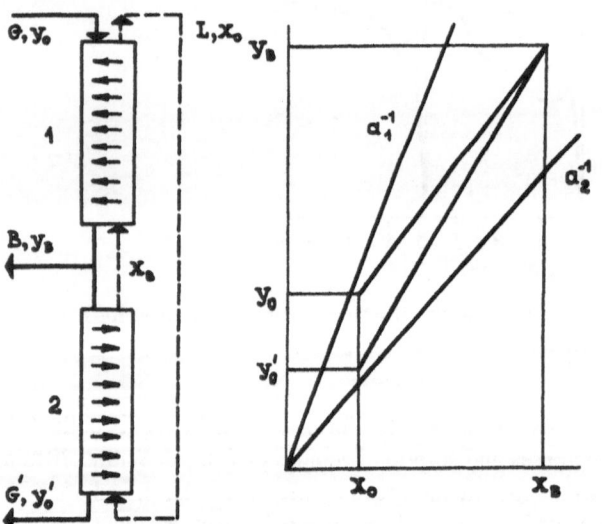

Fig. 5.14. Scheme and x, y-diagram of a two-temperature plant for heavy isotope concentration using a system with positive isotope effect

In the routine two-temperature method the column consists of two parts (hot and cold) and counter-current of gas and liquid with reference to temperature zones is realized. If the solid phase is immobile and the temperature zones are moved in the direction of the second phase movement, the same phase movement with reference to temperature zones can be preserved.

This idea was originally realized for the separation of ions of basic metals in a liquid–ion-exchange system [5.20, 26], and later in the system H_2–Pd. The column filled with palladium sorbent is shown in Fig. 5.15 as a circuit ring. For

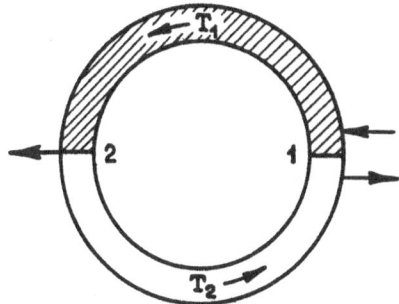

Fig. 5.15. Schematic diagram of the plant with the movement of temperature zones: (1) feed-gas inlet, (2) product-gas outlet. (See text on this page)

concentration of heavy isotopes the feed gas flow enters the cold column (at point 1 between the zones). Hydrogen emerging from the cold zone is the waste. The product is withdrawn at point 2. Feed, withdrawal, and waste can be periodic (once in each period of temperature zone displacement) or continuous but at different column points (in accordance with zone movement). If the process is considered in a system of coordinates moving together with the boundaries of the temperature zones, the theory of the two-temperature method derived for gas–liquid systems [5.19, 24, 25] is applicable to this process. Expressing the velocity of hydrogen and hydride phase flow with reference to the boundaries of the temperature zones, it is possible to find flows of gas G and hydrogen in the hydride phase L:

$$G = (w_g - w_s)S_{fr}\varrho_g ,$$ (5.1)

$$L = w_s a_H S ,$$ (5.2)

where w_g, w_s are the velocities of the gas and temperature zones; S_{fr}, S are free and total cross sections of the column; ϱ_g is the gas density in the column; a_H the sorbent capacity (amount of hydrogen dissolved per unit volume of column with sorbent).

As is noted above, the separation efficiency depends considerably on the flow ratio λ. The function $K = f(\lambda)$ has the sharp maximum at $\lambda^0 = (\alpha_1\alpha_2)^{1/2}$ (at equal numbers of theoretical plates (NTP) of separation in hot and cold zones of the plant).

It follows from (5.1, 2) that

$$\lambda = \frac{G}{L} = \frac{F}{L} - \frac{S_0\varrho_g}{a_H}$$ (5.3)

where S_0 is part of the free section of the column. This equation enables one to relate the feed flow F to the flow of the solid phase L and, hence, to the temperature zone velocity. Furthermore, knowing the separation factors α_1 and α_2 it is possible to find λ^0, the condition under which the greatest separation efficiency is reached in a two-temperature plant with movement of the temperature phases.

Figure 5.16 shows schematically the two-temperature plant with movement of the temperature phases, in which the experiments on separation of the H–D mixture at atmospheric pressure and temperatures $T_1 = 285\,\mathrm{K}$ and $T_2 = 303\,\mathrm{K}$ are carried out. Separation column (1) is a horizontal ring-shaped quartz tube of inner

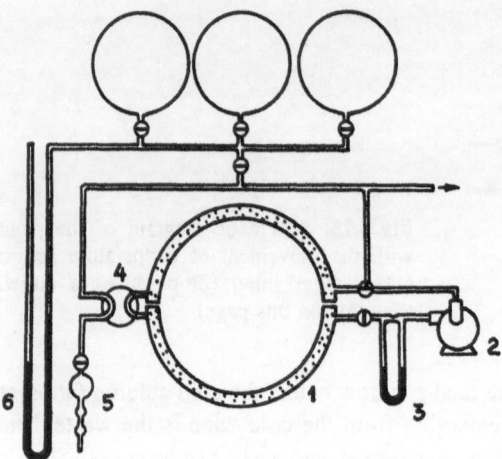

Fig. 5.16. Scheme of a two-temperature plant with continuous movement of temperature zones. (1) separation column, (2) circulatory pump, (3) flowmeter, (4) valve, (5) container for sampling, (6) manometer

diameter 0.3 cm. The column length (including both parts) is equal to 86 cm. The column is filled with granulated Pd powder of grain size 0.5 mm. The mass of palladium is 24 g.

The movement of temperature zones is simplified because the column is ring-shaped and horizontal. The column heating is implemented by means of a nichrome resistance element wound around the column (in a preliminary palladium test the column is heated up to 570 K). The cold zone is created by spraying half the column with cold water; hot water from the thermostat is used to heat the remainder of the column. The movement of temperature zones is achieved by rotation of the spray system. The pump (2) is used for gas circulation and the flowmeter (3) serves to measure the gas velocity. Gas sampling is performed with the use of the cock (4) at the point in time at which the boundaries of the temperature zones coincide with the point of gas sampling. The samples are withdrawn in the container (5), which is further used as a tube in the spectral device. It enables the sample amount to be reduced to 0.5 cm³.

The experiments are carried out in the plant at various flow ratios (changing the temperature zone velocity at constant gas flow being of 0.5 l/h). Figure 5.17 presents the dependence of the separation degree on the flow ratio. The dependence obtained is sharp and characteristic for the two-temperature separation method. During operation samples are withdrawn at the "poor" and "rich" ends of the column. The steady separation level shown in Fig. 5.17 is reached in no more than 4–6 h.

Although the experiments are carried out with only slight difference in temperatures between hot and cold zones, on account of the high separation efficiency, the change of isotope concentration appears to be considerable (from 2.5 to 30 at% of deuterium).

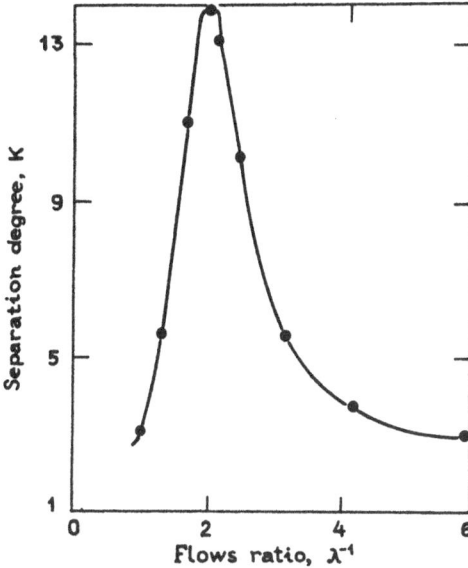

Fig. 5.17. Dependence of the separation degree for the H–D mixture on the flows ratio

5.2.4 Optimum Conditions for separation

The data on isotope equilibrium and efficiency of interphase isotope exchange presented in the previous sections allow one to estimate the optimum parameters of continuous counter-current separation of hydrogen isotopes in the systems H_2–Pd and H_2–LaNi$_5$. Since the mass-transfer characteristics, strictly speaking, depend on construction and sizes of separation column and on sorbent characteristics (including geometrical ones), the approach under consideration and the resulting conclusions are more important than final results of optimization.

Since systems with hydride phases of metals or IMC are appropriate for tackling relatively low-scale separation tasks (due to their lower productivity in comparison to separation processes in gas (vapour)–liquid systems), we consider not the most economical two-temperature scheme but a simpler separation method in a column with flow conversion. Let us determine the optimum temperature in the separation column (at which the volume is minimal) filled with granulated sorbent based on LaNi$_5$, i. e., using a system with a positive isotope effect.

The height of the sorbent bed in the column depends, apart from on α, on the required separation degree K, the value of relative withdrawal θ, and HTU h_{0g} [5.19, 25]:

$$H_c = h_{0g} N_{0g} = h_{0g} \frac{\alpha}{\alpha - 1} \ln \left(\frac{K - \theta}{1 - \theta} \right) \tag{5.4}$$

where N_{0g} is the number of transfer units (NTU) expressed in terms of the driving force for the gas phase.

Deriving HTU from (4.68) we obtain

$$H_c = G_{sp}\left(\frac{1}{\beta_p a_{gr}} + \frac{1}{\alpha\beta_m a_{gr}}\right)\alpha\ln\left(\frac{K-\theta}{1-\theta}\right)/(\alpha-1).\tag{5.5}$$

In the plant of productivity P fed by source flow with isotope content y_0 the source flow depends on the extraction degree ($E = \theta E_m$):

$$G = \frac{P}{y_0\theta E_m},\tag{5.6}$$

where E_m is the maximal extraction degree (at $y_0 \ll 1$ $E_m = 1-1/\alpha$). The column section is expressed as:

$$S_k = \frac{G}{G_{sp}} = \frac{P}{G_{sp}y_0\theta(1 - 1/\alpha)}.\tag{5.7}$$

From (5.5, 7) we find that the column volume is equal to

$$V_k = \frac{P}{y_0\theta}\ln\left(\frac{K-\theta}{1-\theta}\right)\left[\left(\frac{\alpha}{\alpha-1}\right)^2\left(\frac{1}{\beta_p a_{gr}} + \frac{1}{\alpha\beta_m a_{gr}}\right)\right].\tag{5.8}$$

At definite values of P, x_0, K, and θ the temperature dependence of V_k is determined by the factor

$$\left(\frac{\alpha}{\alpha-1}\right)^2\left(\frac{1}{\beta_p a_{gr}} + \frac{1}{\alpha\beta_m a_{gr}}\right)$$

Since this system is not appropriate for use at the stage of final concentration (due to decrease of α), the calculations are performed for the range of initial concentration. As is evident from Fig. 5.18, the optimum temperatures for H–T and H–D separation are close and lie in the range 230–250 K.

In the H$_2$–Pd system, the temperature dependence of the column volume is weaker than in the H$_2$–LaNi$_5$ system. The optimum position depends on the size of the sorbent grains and ranges from 270 to 300 K independent of the mixture being separated [5.20].

Let us dwell on the question of what column load should be choosen. If the impairment of separation efficiency due to axial dispersion in the gas phase is neglected, the volume of the separation part of column remains constant at any loading because of the linear character of HTU dependence on loading. In this case the load should be increased, since the lower the column diameter, the lower the adverse influence of lateral irregularity.

To determine the allowable load in the column by the method of response curves (C-curves), the axial dispersion during hydrogen passage through a column with an immobile bed of different granulated sorbents was studied [5.27]. The experiments showed that a considerable change in temperature (in the range 77–300 K) does not essentially affect the axial dispersion. When the column is filled with grains of size 0.3–1 mm, the gas velocity may not exceed 0.2 m/s. In plants

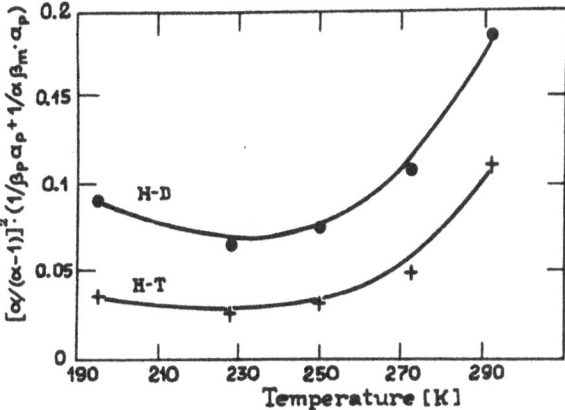

Fig. 5.18. Influence of temperature on the volume of the counter-current separation column with the use of granulated sorbent on the basis of LaNi$_5$ at $P_{H_2} = 1.9$ MPa

of high productivity, apparently, multitube columns are appropriate for retaining low values of HTU.

Flow-conversion units are important elements of any isotope-separation plant. Maintenance of practically 100 % efficiency of flow conversion is the main task of the units. Incompleteness of flow conversion at the column end enriched with the desired product results in a loss of product. If the losses become comparable to the value of the maximal allowable withdrawal, the column does not produce any product. The flow-conversion unit efficiency when using systems with hydride phases depends primarily on the character of the phase diagram H$_2$–solid phase and on the kinetics of sorption-desorption processes in the system with the hydride phase. The degree of thermal desorption is also determined by the temperature regime in the desorber; a large surface for heat transfer is the principal condition for its effective operation.

The sorption zone width depends on the gas velocity, the kinetics of sorption (hydrogenation) and the temperature regime. As shown by the experiments performed on the dynamics of the sorbent based on Pd and LaNi$_5$ hydrogenation, this process proceeds fast and independently of temperature. It is controlled by molecular hydrogen diffusion in pores of sorbent grains (provided the heat of hydride formation is removed effectively). Thus, temperature only weakly affects the width of the sorption front in comparison with the efficiency of interphase isotope exchange. Figures 5.19, 20 present the results of experiments on the dynamics of sorbents based on Pd and LaNi$_5$ hydrogenation obtained by passing the mixture of hydrogen with argon (1 : 1) through the column and by determination of the hydrogen content in the outlet gas by the method of heat conduction [5.19].

If the pressure of hydride formation is significantly lower than the operating pressure in the column, a change of the latter (in case of a molecular mechanism of hydrogen diffusion in pores of the grains) may not significantly affect the width of the sorption front. If hydrogen diffusion in the pores is by the Knudsen mechanism,

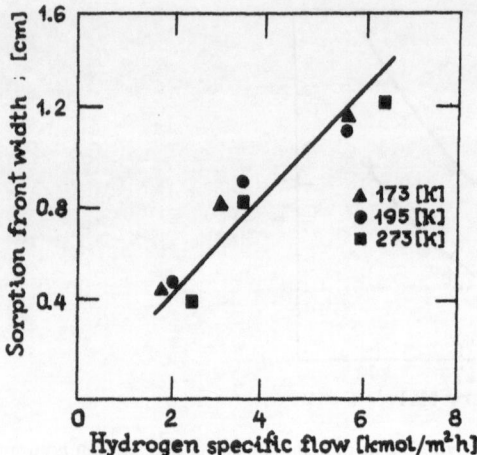

Fig. 5.19. Dependence of the width of the hydrogen sorption front on the specific hydrogen flow in the column of diameter 1 cm, filled with granulated palladium sorbent (grain size – 1.3 mm), at $P = 0.05$ MPa

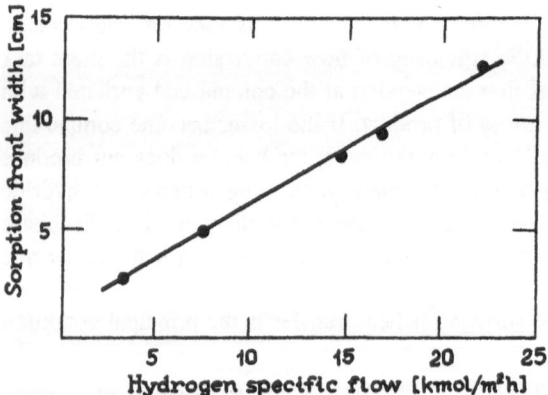

Fig. 5.20. Dependence of the width of the sorption front on specific hydrogen flow in the column of diameter 1 cm, filled with granulated sorbent on the basis of LaNi₅ (grain size – 1.5 mm), at $P = 0.5$ MPa and $T = 273$ K

pressure elevation enhances the mass-transfer. However, the width of the sorption front can also depend on the axial diffusion effect, which should be taken into account when calculating the upper flow-conversion unit.

When comparing different working substances, in addition to the above-considered characteristics (separation factors, HETP), one should also consider sorbent properties such as capacity for hydrogen (more exactly, amount of hydrogen per unit sorbent volume) and heat of hydrogen sorption (hydride formation) ΔH_H, which affects the energy consumption in flow conversion.

The basic characteristics of those H₂–Me(IMC) systems, for which the data on the efficiency of the mass-transfer are available are presented in Table 5.6. The

H_2–Pd system is characterized by the largest thermodynamic isotope effect. The system H_2–LaNi$_5$ is one of the best studied. Substitution of La by a natural mixture of elements of the lanthanum row does not change the thermodynamic and mass-transfer characteristics of the system but makes this compound less expensive and more readily available (manufactured in a number of countries). The compound $Ti_{0.8}Zr_{0.2}CrMn$, one of the promising class of compounds of the Laves-phase-type, is characterized by a rather high isotope effect, but the required hydrogen capacity and high mass-transfer rate are reached only at elevated pressure.

Table 5.6. Comparison of different systems working with solid phase

Sorbent	ΔH_H	Density of granulated layer	Capacity	T	α_{HT}	P	HETP at $G_{sp} = 1$
	kJ/mol	g/cm^3	cm^{3}H$_2$/g	K		MPa	mol/m^{2}s
Pd*	41	1.4	50	298	1/2.7	0.1	0.3–0.4
LaNi$_5$**	29	2.0	150	250	1.4	0.5	1.0***
Ti$_{0.8}$Zr$_{0.2}$CrMn**	–	2.0	–	250	2.0	1.9	1.1***
Zeolite NaX	8.4	0.75	120	77	2.0	0.1	0.3–0.4

* granulated sorbent (grain size $\approx$ 1 mm) containing 75 mass% of Pd
** granulated sorbent (spherical grains of size 1.0–1.5 mm) with polymeric binder in amounts of 20 mass%
*** at $G_{sp} = 3$ mol/m^{2}s

For comparison, Table 5.6 contains the characteristics of the H_2–zeolite NaX system, with the use of which experiments on the separation of the isotope mixture H–T were carried out in the above-considered plant with sectioned column in conditions similar to the experiments with the palladium-based sorbent.

The hydrogen–palladium system is characterized by the largest separation factors of all binary mixtures of hydrogen isotopes and by the possibility of achieving separation at room temperature. It is useful to note once again that this system has an important advantage in terms of its use in the final tritium concentration, namely, the increase of α^{-1} with increase in heavy isotope concentration. IMC are significantly cheaper than palladium but are characterized by a lower rate of isotope exchange and markedly lower separation factors. Hence, reduced temperature and elevated pressure are required for their effective use. Finally, zeolites are the cheapest widely manufactured industrially in granulated form. However, low temperature (77 K) is required for plant operation, which complicates the equipment. It should be noted that despite such low temperature, zeolites are characterized by a high efficiency of the interphase isotope exchange, since the steps of molecular dissociation and implantation of atoms into the crystal lattice (slow at low temperatures) are absent. At the same time, dissociation of hydrogen molecules in hydride-forming metals and IMC is followed by the HMIE reactions, which make possible pure isotope extraction in a single column. In the case of zeolites this reaction must be performed at elevated temperature in a special reactor with a catalyst, similar to the low-temperature rectification of hydrogen.

5.3 Use of Systems with a Hydride Phase
for Practical Separation
of Tritium-Containing Isotope Mixtures

The separation of tritium-containing mixtures using hydrogen–solid phase systems differs from the majority of other methods used for heavy-water production in that the operation substance are not prone to radiolysis. Due to the lower toxicity of gaseous tritium [5.28] the use of hydrogen (but not water as applied in the GS-process and in the H$_2$–H$_2$O system, considered as promising) simplifies the solution of radiation safety problems. However, the comparatively low passability of columns with a solid phase limits their potential for use in small-scale tasks. Consider such problems as the purification of a heavy-water moderator of a nuclear reactor and the regeneration of the D–T mixture emerging from the plasma chamber of a thermonuclear reactor (TNR). These are considered to be difficult but do not require processing of high flow rates.

Estimating systems with IMC as promising we will restrict our consideration to the system with Pd. Based on this system chromatographic plants for preliminary enrichment of analytical tritium samples during radiometric check in the environment were tested and chromatographic variants of systems for hydrogen isotope separation in TNR the fuel cycle were suggested [5.8–10].

Purification of the heavy-water moderator of a nuclear reactor using a sectioned column filled with Pd sorbent is considered in [5.19]. Figure 5.21 presents the scheme of the plant for isotope purification including the separation column for tritium and protium extraction and the unit for catalytic isotope exchange (CIE) between water and hydrogen. We do not dwell on the CIE unit, which, as in the method of low-temperature rectification of deuterium now in use [5.29], serves to convert tritium and protium from heavy water to deuterium circulating in a closed circuit.

Consider the stage of isotope separation: Deuterium from the CIE unit is fed into the column for tritium concentration, where the following isotope exchange reaction proceeds:

$$D_2 + T(Pd) \leftrightarrow DT + D(Pd) .\tag{5.9}$$

The temperature dependence of the separation factor for this reaction is expressed by

$$\ln \alpha_{DT}^{-1} = 0.0023 + \frac{113.7}{T} .\tag{5.10}$$

Gas enriched with tritium is withdrawn from the upper part of the column. In principle, due to the D$_2$+T$_2$ $\leftrightarrow$ 2DT reaction, concentration can be performed up to the commercial product. A second column of smaller section is desirable (II stage) to decrease tritium hold-up in the separation equipment for final concentration (at this stage the separation factor increases to $\alpha_{TD}^{-1} = 1.53$ at 298 K).

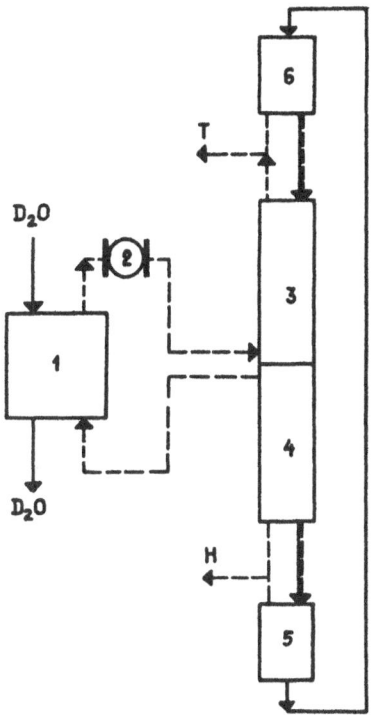

Fig. 5.21. Scheme of the plant for isotope purification of heavy water [5.19]: (1) CIE unit, (2) circulatory compressor, (3) column for tritium concentration, (4) column for protium concentration, (5) desorber, (6) absorber

The lower column serves to concentrate protium, which is withdrawn from the gas flow emerging from the desorber in the form of an equilibrium mixture of $H_2 + HD + D_2$.

To compare the method under consideration with the method of low-temperature rectification of deuterium, Table 5.7 presents the results of a calculation for a plant of the same productivity as that which purifies the heavy-water moderator of the reactor in Grenoble [5.29]. The table shows that the characteristic of the basic column allow initial tritium concentration (up to 5.55×10^{13} Bk/l converting to heavy water) and also protium concentration up to 40 at%.

Isotope purification using the H_2–Pd system has the following advantages:

1) the separation proceeds at room temperature and atmospheric pressure;
2) final tritium concentration can take place in the separation apparatus without a special stage of catalytic decomposition of HD in a separate reactor;
3) low energy consumption in isotope purification.

The energy consumption is determined purely by the necessity to heat up the palladium sorbent (in hydride form) to the temperature of the desorber (420–470 K) and supply the heat of desorption. Since no spezial coolant (e. g., water) is required for cooling the gas emerging from the desorber and Pd sorbent, the corresponding energy consumption is insignificant. However, the high price of palladium is a

Table 5.7. Basic characteristics of the plant for isotope purification of heavy water of productivity by tritium 8.9×10^6 GBk/year (at a tritium concentration in the source flow of 70 GBk/l) and by protium $100 \, l \, H_2O$/year (at a protium concentration in the source flow of 0.4 at%)

Characteristic	Plant with palladium sorbent	Plant of low-temperature rectification in Grenoble
Temperature [K]	298	24–25
α_{DH}	1/2.47	1.51
α_{DT}	1/1.47	1.22
NTP	30	110
Column section/cm^2	1400*	490

* at $G_{sp} = 5.55 \, mol/m^2 s$

basic drawback of this method and the search for other effective systems is an urgent problem.

The method of low-temperature rectification of hydrogen was suggested for regeneration of the D–T mixture emerging from the plasma chamber of TNR. The study [5.30] considers the variants of schemes of low-temperature rectification including from four to six separation columns (the overall NTP is 200–300) and two reactors for the performance of the HMIE reaction at 300 K. The energy consumption for separation is more than 10 kW for the regeneration of 22 g-mol/h of the mixture on account of the low operating temperature in the columns (23–25 K) and the repeated heating and cooling of the gas flow. Its considerable tritium content poses a problem of disposal for the waste hydrogen flow.

The work [5.22] considers another variant of the same task (regeneration of 22 g-mol/h of the D–T mixture by purification of protium to a residual content of 1 at%) and shows the possibility of effective purification of protium in a small diameter column (of volume less than 2 l).

At present a universal system is being explored for the separation of hydrogen isotopes applicable to tritium processing both in water and in gas. In particular, the project Isotope Separation System (ISS) deals with a scheme [5.31] which includes two columns for water rectification (they serve to remove tritium from water flows) and four columns for low-temperature rectification: the first column works in the range of low tritium content and serves to remove tritium from the waste hydrogen flow; the second one works at intermediate tritium concentrations; the third produces pure deuterium (99.99 at%); and the fourth completely purifies the D–T mixture of protium (the final product is a D–T mixture of composition: 80 at% tritium + 20 at% deuterium + about 10^{-3} at% of protium). All separation columns are connected to one another by gas flows: columns for water rectification (by the CIE units) and columns 2–4 (by the HMIE reactors) are connected with column 1. The scheme includes 5 reactors for HMIE (two HMIE reactors are required for operation of columns 1 and 2). ISS has a significant drawback, apart from the complicated scheme, namely, hydrogen rectification requires extremely low temperature and special equipment for a complete purification of the gas. The computed tritium content in ISS exceeds 300 g bringing about a lot of problems due to the high radioactivity.

Since isotope separation on palladium is followed by the HMIE reactions, ISS can be significantly simplified and the tritium content in it can be reduced. Let us illustrate this fact by the example of the D–T mixture purification of protium, which makes the main contribution ($\approx 70\%$) to the overall amount of tritium in the separation system.

ISS of ITER purifies 71.4 mol/h of the D–T mixture (1.0 at% of hydrogen + 49.5 at% of deuterium + 49.5 at% of tritium) of protium and simultaneously concentrates protium and deuterium at another end of the column (at a tritium content of ≈ 2 at%). When using the H_2–Pd system, a single column is sufficient to solve this problem. It operates according to the scheme shown in Fig. 5.5a. Considering the mass balance equation the withdrawal flows B (regenerated D–T mixture) and P (protium and deuterium) and the isotopic composition of flow P can be calculated, starting with the feed flow F and its isotopic composition. The values of B and P are presented in Table 5.8. The isotope composition of the flow P is the following: 2.56 at% of hydrogen + 95.44 at% of deuterium + 2 at% of tritium.

Table 5.8. Basic characteristics of the plant for regeneration of 71.4 g mol/h of the D–T mixture emerging from the plasma chamber of TNR with the use of palladium sorbent in a sectioned column

Characteristic	The upper part of the column	The lower part of the column
Withdrawal flow [g mol/h]	$B = 43.5$	$P = 27.9$
Gas flow, G [g mol/h]	197.6	126.2
Flows ratio, λ	1.28	0.82
Separation factor	$\alpha_H^{-1} = 2.4$	$\alpha_{DT}^{-1} = 1.47$
NTP	11	18
Number of separation sections	4	6
Tritium amount [g]	24	10

The upper part of the column ensures purification of protium from 1 to 10^{-3} at% ($K_H = 1000$). Since the ratio of tritium and deuterium is equal to 1 in the source mixture, when calculating this part of the column, the value $\alpha_H^{-1} = 2.4$ (found from a relation similar to (3.72) at $\alpha_{HD}^{-1} = 2.01$ and $\alpha_{HT}^{-1} = 2.67$) is used. In fact the average separation factor α_H^{-1} is slightly higher due to the increase of this ratio to the value $0.8:0.2 = 4$. At a value of relative withdrawal of $\theta = 0.8$ the flow ratio turns out to be $\lambda = G/L = 1.28$ and NTP, computed by the "plate to plate" method is found equal to 11. As shown by the calculation, it takes nearly 10 theoretical plates of separation at this flow ratio and a separation factor $(\alpha_{D-T}^0)^{-1} = 1.51$ for tritium to be concentrated from 49.5 to 80 at% ($K_T = 4.08$).

In the lower part of the column gas purification of tritium is the most difficult task ($K_T = (49.5 \times 98)/(50.5 \times 2) = 48$). Assuming as value of the separation factor for the D–T mixture $\alpha_{DT}^{-1} = 1.47$ at $\theta = 0.8$ we obtain $\lambda = 0.82$ and using the "plate to plate" method we find NTP = 18. At a specific flow $G_{sp} = 5.55$ g-mol/m²s in the upper part, typified by the largest gas flow, the column section is equal to 98.8 cm².

If the height of the separation section is assumed to be equal to 6 cm, at HETP = 2 cm the upper part of the column includes 4 separation sections and the lower part 6. Thus the dimensions of the column with palladium sorbent appear to be the following (in terms of two sections working as flow-conversion units): height 72 cm, diameter 11 cm. At a density of the granulated Pd sorbent of 1.4 (Table 5.6) and a palladium content of 75% it takes only a moderate amount of palladium (≈ 10 kg).

Table 5.8 also presents the results of calculating the tritium amount in the column in terms of its concentration profile along the column height, whereby the palladium gas capacity being equal to 50 cm³s.t.p./gPd and the sorption section (absorber) is assumed to be filled with a mixture of composition 80 at% of T + 20 at% of D.

The calculation shows that it is possible to solve the problem of regeneration of the D–T mixtures emerging from the plasma chamber of TNR by a compact plant (total column volume 7.1 l) of simple construction (working at room temperature and atmospheric pressure) and using rather small amounts of palladium. Thus, the amount of tritium in the system for isotope separation is reduced by a factor of two.

The results of gas-chromatographic regeneration of the D–T mixture used in the JET project indicate lower efficiency in comparison with the sectioned column. In that case it takes 4 chromatographic columns, each of volume 5 l, to process the D–T mixture in quantities of 20 g-mol/day [5.8, 9]. However, tests for a wide range of tritium concentrations were carried out in such a system of chromatographic columns [5.9].

In conclusion we note that, apart from the above-considered problems, the use of the H_2–Pd system is promising both for environmental protection against tritium discharge at NPS, plants for irradiated nuclear fuel processing, and for the case of emergency at a TNR. Because of the significant flows processed and their relatively low tritium content it is appropriate to use not low-temperature hydrogen rectification but more economical methods, which do not require the tritium conversion from water to hydrogen (GS-process, chemical isotope exchange H_2O–H_2). In these cases a waste-free process can be realized if the tritium concentration process up to the commercial product is performed in a sectioned plant with a palladium sorbent. Such combined schemes for tritium trapping at facilities of the nuclear power industry are considered in [5.19, 33, 34] and for a radiation safety system in [5.35]. In addition, the final concentration of tritium in the sectioned column with a palladium sorbent was considered [5.22, 23] (for water used as heat-transport substance in a TNR and purified by the method of chemical isotope exchange in the system H_2O-H_2 or using the GS-process) and protium removal from tritium extracted from a blanket was also discussed. Evidently, if palladium is substituted by other cheaper metals or IMC, the field of application of systems with hydride phases for the separation of tritium-containing isotope mixtures will considerably expand.

References

Chapter 1

1.1 G. Alefeld, J. Völkl (eds.): *Hydrogen in Metals* Vols. 1 and 2 (Springer, Berlin, Heidelberg 1978)
1.2 L. Schlapbach (ed.): *Hydrogen in Intermetallic Compounds* Vol. 1 (Springer, Berlin, Heidelberg 1988)
1.3 L. Schlapbach (ed.): *Hydrogen in Intermetallic Compounds* Vol. 2 (Springer, Berlin, Heidelberg 1992)

Chapter 2

2.1 T. B. Flanagan, W. A. Oates: Ber. Buns. Ges. **76** 706 (1972)
2.2 G. A. Ferguson, A. I. Schindler, T. Tanaka, T. Morita: Phys. Rev., Ser 2A **137**, 483 (1965)
2.3 E. O. Wollan, J. W. Cable, W. C. Koehler: J. Phys. Chem. Solids **24**, 1141 (1963)
2.4 W. E. Wallace: J. Chem. Phys. **35**, 2156 (1961)
2.5 S. S. Sidhu, L. Heaton, M. U. Mueller: J. Appl. Phys. **30**, 1323 (1959)
2.6 P. T. Gallagher, W. A. Oates: Trans. Metall. Soc-AIME **245**, 179 (1969)
2.7 P. S. Rudman, G. Sandrock: Ann. Rev. Mater. Sci. **12**, 271 (1982)
2.8 R. Griessen, T. Riesterer: "Heat of Formation Models" in Topics Appl. Phys. Vol. 63 (Springer, Berlin, Heidelberg 1988) p. 220
2.9 W. Drexel, A. Murani, D. Tocchetti, W. Kley, I. Sosnowska, D. K. Ross: J. Phys. Chem. Sol. **37**, 1135 (1976)
2.10 D. G. Hunt, D. K. Ross: J. Less-Common Met. **49**, 169 (1976)
2.11 R. Lässer, K-H. Klatt: Phys. Rev. B **28**, 748 (1983)
2.12 R. Lässer: Z. Phys. Chem. NF **143**, 23 (1985)
2.13 V. Trentin, Ph. Brossard, D. Schweich: Chem. Eng. Sci. **48**, 873 (1993)
2.14 E. Wicke, H. Brodowsky: "Hydrogen in Palladium and Palladium Alloys" in Topics Appl. Phys., Vol. 29 (Springer, Berlin, Heidelber 1978) p. 73
2.15 A. Biris, R. V. Buccur, P. Ghete, E. Indrea, D. Lupu: J. Less-Common Met. **49**, 477 (1976)
2.16 B. M. Andreev, O. V. Dobryanin, E. P. Magomedbekov, Yu. S. Pak, V. V. Shitikov: Zh. Fiz. Khim. **56**, 463 (1982) (in Russian)
2.17 E. Wicke, G. Nernst: Ber. Bunsenges. Phys. Chem. **68**, 224 (1964)
2.18 H. Brodowsky, E. Poeschel: Z. Phys. Chem. NF **44**, 143 (1965)
2.19 S. Schmidt, G. Sicking: Z. Naturforsch. **33a**, 1328 (1978)
2.20 P. Mecking: Ber. KFA Jülich **1779** (1982)
2.21 H. P. Bleichert: Ber. KFA Jülich **2005** (1985)
2.22 P. Meuffels: Ber. KFA Jülich **2081** (1986)

2.23 J. A. Pryde, I. S. T. Tsong: Trans. Faraday Soc. **67**, 297 (1971)
2.24 P. Dantzer, O. J. Kleppa, M. E. Melnichak: J. Chem. Phys. **64**, 139 (1976)
2.25 C. Boffito, B. Ferrario, D. Martelli: J. Vac. Sci. Technol. **A1**, 1279 (1983)
2.26 F. Ricca: J. Phys. Chem. **71**, 3632 (1967)
2.27 K. Ichimura, M. Matsuyama, K. Watanabe, T. Takeuchi: J. Vac. Sci. Technol. **A6**, 2541 (1988)
2.28 S. A. Steward: J. Chem. Phys. **63**, 975 (1975)
2.29 D. H. W. Carstens, W. R. David: *Isotopic Effects in Hydrides of Lanthanum-Nickel Alloys*, ed. by T. N. Veziroglu (Pergamon, Oxford 1982) p. 477
2.30 J. F. Lynch, J. R. Johnson, J. J. Reilly: Z. Phys. Chem. NF **117**, 229 (1979)
2.31 J. F. Lynch, J. J. Reilly, J. Tanaka: "The Titanium–Molybdenum–Hydrogen System" in Advances in Chemistry Series, Vol. 167 (American Chem. Soc., Washington D.C. 1978) p. 342
2.32 R. H. Wiswall, J. J. Reilly: Inorg. Chem. **11**, 1691 (1972)
2.33 H. Wenzl: Int. Met. Rev. **27**, 140 (1982)
2.34 T. Tanabe, S. Mirua, S. Imoto: J. Nucl. Sci. Technol. **16**, 690 (1979)
2.35 B. Andreev, E. Magomedbekov, A. Shafiev, V. Shitikov: J. Less-Common Met. **90**, 161 (1983)
2.36 B. M. Andreev, E. P. Magomedbekov, V. V. Shitikov: At. Energ. **55**, 102 (1983) (in Russian)
2.37 B. M. Andreev, E. P. Magomedbekov, Yu. S. Pak, M. G. Zagliev: Zh. Fiz. Khim. **58**, 2841 (1984) (in Russian)
2.38 G. D. Sandrock, J. J. Murray, M. L. Post, J. B. Taylor: Mat. Res. Bull. **17**, 887 (1982)
2.39 M. Devillers, M. Sirch, R. D. Penzhorn: Z. Phys. Chem. NF **164**, 1355 (1989)
2.40 M. E. Kost, L. N. Padurets, A. A. Chertkov, V. I. Mikheeva: Zh. Neorg. Khim. **25**, 471 (1980) (in Russian)
2.41 K. Nakamura, T. Hoshi: J. Vac. Technol. **A3**, 34 (1985)
2.42 O. J. Kleppa, M. E. Melnichak, T. V. Charlu: J. Chem. Thermodyn. **5**, 595 (1973)
2.43 O. J. Kleppa, G. Boureau: J. Chem. Thermodyn. **9**, 543 (1977)
2.44 G. Picard, O. J. Kleppa, G. Boureau: J. Chem. Phys. **69**, 5549 (1978)
2.45 O. J. Kleppa: Ber. Bunsenges. Phys. Chem. **87**, 741 (1983)
2.46 J. J. Murray, M. L. Post, D. M. Crant: Z. Phys. Chem. **163**, 135 (1989)
2.47 T. B. Flanagan, B. S. Bowerman, G. E. Biehl: Scripta Metall. **14**, 443 (1980)
2.48 B. S. Bowerman, C. A. Wulff, G. E. Biehl, T. B. Flanagan: J. Less-Common Met. **73**, 1 (1980)
2.49 P. Dantzer, E. Orgaz, V. K. Sihha: Z. Phys. Chem. **163**, 141 (1989)
2.50 A. V. Krupentshenko, E. P. Magomedbekov, I. I. Wedernikova: Zh. Fiz. Khim. **54**, 2897 (1990) (in Russian)
2.51 A. V. Krupentshenko, E. P. Magomedbekov, B. M. Andreev, O. W. Dyatlova: Zh. Neorg. Khim. **36**, 175 (1991) (English transl.: Russ. J. Inorg. Chem. **36**, 98 (1991))
2.52 A. V. Krupentshenko, E. P. Magomedbekov, B. M. Andreev: Zh. Neorg. Khim. **36**, 1595 (1991) (English transl.: Russ. J. Inorg. Chem. **36**, 907 (1991))
2.53 A. V. Krupentshenko, E. P. Magomedbekov: Zh. Neorg. Khim. **37**, 174 (1992) (in Russian)
2.54 A. N. Perevesenzev, E. Lanzel, O. J. Eder: J. Less-Common Met. **143**, 39 (1988)
2.55 A. Pebler, E. A. Gulbransen: Trans. AIME **239**, 1593 (1967)
2.56 V. K. Sinha, F. Pourarian, W. E. Wallace: J. Less-Common Met. **87**, 283 (1982)
2.57 F. Pourarian, V. K. Sinha, W. E. Wallace: J. Less-Common Met. **96**, 237 (1984)
2.58 T. Riesterer, P. Koefel, L. Schlapbach: J. Less-Common Met. **101**, 221 (1984)
2.59 T. B. Flanagan, C. A. Wulff, B. S. Bowerman: J. Solid State Chem. **34**, 215 (1980)
2.60 G. Pfeiffer, H. Wipf: J. Phys. F: Metal phys. **6**, 167 (1976)
2.61 H. Oesterreicher: J. Phys. Chem. **85**, 2319 (1981)
2.62 E. P. Magomedbekov, B. M. Andreev, A. V. Korolev: Zh. Fiz. Khim. **64**, 434 (1980) (in Russian)

2.63 I. I. Wedernikova, E. P. Magomedbekov, B. M. Andreev, A. V. Krupentchenko, A. V. Korolev: Zh. Fiz. Khim. **65**, 1657 (1991) (in Russian)
2.64 H. H. Van Mal, K. H. Buschow, A. R. Miedema: J. Less-Common Met. **35**, 65 (1974)
2.65 D. Shaltiel: J. Less-Common Met. **62**, 407 (1978)
2.66 E. P. Magomedbekov, A. V. Krupentshenko: Zh. Neorg. Khim. **38**, 1732 (1993) (in Russian)
2.67 E. Magomedbekov, A. Krupentshenko, G. Sicking: J. Alloys and Compounds **199**, 73 (1993)

Chapter 3

3.1 P. W. Albers, G. H. Sicking, D. K. Ross: J. Phys. Condens. Matter **1**, 6025 (1989)
3.2 J. J. Rush, J. M. Rowe, D. Richter: Z. Phys. B – Condens. Matter **55**, 283 (1984)
3.3 D. Basmadjian: Can. J. Chem. **38**, 149 (1960)
3.4 W. E. Kochurikhin, Ya. D. Zelvenskii: Zh. Fiz. Khim. **38**, 2594 (1964) (in Russian)
3.5 B. M. Andreev, M. M. Domanov, G. I. Medvedeva: Isotopenpraxis **6**, 236 (1971)
3.6 B. Andreev, V. Shitikov, E. Magomedbekov, A. Shafiev: J. Less-Common Met. **90**, 161 (1983)
3.7 B. M. Andreev, O. V. Dobryanin, E. P. Magomedbekov, Yu. S. Pak, V. V. Shitikov: Zh. Fiz. Khim. **56**, 463 (1982) (in Russian)
3.8 B. M. Andreev, E. P. Magomedbekov, A. S. Polevoi: Tr. Mosk. Khim.-Tekhnol. Inst. im. D.I. Mendeleeva **130**, 45 (1984) (in Russian)
3.9 R. H. Wiswall, J. J. Reily: Inorg. Chem. **11**, 1691 (1972)
3.10 S. Imoto, T. Tanaba, K. Utsunomiya: Int. J. Hydr. Energy **7**, 597 (1982)
3.11 M. Karas, E. P. Magomedbekov, G. H. Sicking: J. Less-Common Met. **159**, 307 (1990)
3.12 B. M. Andreev, A. S. Polevoi, A. N. Perevezentsev: At. Energy **45**, 53 (1978) (in Russian)
3.13 G. Sicking, E. Magomedbekov, R. Hempelmann: Ber. Bunsenges. Phys. Chem. **85**, 686 (1981)
3.14 A. I. Brodskii: *Chemistry of Isotopes* (Akad. Nauk SSSR, Moscow 1957) pp. 1–594 (in Russian)
3.15 B. M. Andreev, Ya. D. Zelvenskii, S. G. Katalnikov: *Heavy Hydrogen Isotopes in Nuclear Technology* (Energoatomisdat. Moscow 1987) pp. 1–456 (in Russian)
3.16 J. Bron, C. F. Chang, M. Wolfsberg: Z. Naturforschung **28a**, 129 (1972)
3.17 G. Sicking: Z. Phys. Chem. NF **93**, 53 (1974)
3.18 G. Sicking, E. Magomedbekov: *Met.-Hydrogen Syste.*, Proc. Miami Int. Symp., Miami Beach, Florida, USA, April 13–15, 1981, ed. by T. N. Veziroglu (Pergamon, Oxford 1982) pp. 71–88
3.19 B. M. Andreev, E. P. Magomedbekov, V. V. Shitikov: At. Energy **55**, 102 (1983) (English transl.: Sov. At. Energy **55**, 535 (1983))
3.20 G. Sicking: J. Less-Common Met. **101**, 169 (1984)
3.21 B. M. Andreev, G. H. Sicking: Ber. Bunsenges. Phys. Chem. **91**, 177 (1987)
3.22 B. M. Andreev, A. N. Perevezentsev, I. L. Selivanenko: *Isotope Separation and Chemical Exchange Uranium Enrichment* Proc. Int. Symp., Tokio, Japan, October 29–November 1, 1990, ed. by Y. Fujii, T. Ishida, K. Takeuchi, Bull. Research Lab. Nucl. Reactors **1**, 1 (1992)
3.23 Ya. M. Warshawskii, S. E. Waysberg: Zh. Fiz. Khim. **29**, 523 (1955) (in Russian)
3.24 V. S. Parbuzin, N. I. Malyavskii: Zh. Fiz. Khim. **50**, 2944 (1976) (in Russian)
3.25 N. I. Malyavskii, V. S. Parbuzin: Vestn. Mosk. Univ., Ser. Khim. **18**, 111 (1977) (in Russian)

3.26 B. M. Andreev, Ya. D. Zelvenskii, S. G. Katalnikov, W. W. Uborskii: Isotopenpraxis **12**, 440 (1977) (in Russian)

3.27 F. Botter: J. Phys. Chem. **69**, 2485 (1965)

3.28 F. Botter: J. Less-Common Met. **49**, 111 (1976)

3.29 E. Wicke, G. Nernst: Ber. Bunsenges. Phys. Chem. **68**, 224 (1964)

3.30 M. N. Domanov, B. M. Andreev, S. E. Gilburd: Tr. Mosk. Khim.-Tekhnol. Inst. im. D.I. Mendeleeva **67**, 104 (1970) (in Russian)

3.31 F. Botter: *Isotope Effects in Physical and Chemical Processes* Int. Meeting, Cluj. Roumania, June 25–28, 1973

3.32 G. Sicking: Ber. Bunsenges. Phys. Chem. **76**, 790 (1972)

3.33 B. M. Andreev, M. M. Domanov: Zh. Fiz. Khim. **49**, 1258 (1975) (in Russian)

3.34 B. M. Andreev, A. N. Perevezentsev, I. A. Mandrykin, N. F. Myasoedov: Radiokhimia **28**, 212 (1986) (English transl.: Sov. Radiochem. **28**, 188 (1986))

3.35 T. Tanabe, S. Miura, S. Imoto: J. Nucl. Sci. Technol. **16**, 690 (1979)

3.36 B. M. Andreev, E. P. Magomedbekov, V. V. Shititkov: Zh. Fiz. Khim. **58**, 2418 (1984) (in Russian)

3.37 A. S. Horen, M. W. Lee: Fus. Technol. **21**, 282 (1992)

3.38 T. R. P. Gibb: Adv. Chem. Ser. **39**, 99 (1992)

3.39 G. G. Libowitz: *The Solid State Chemistry of Binary Metal Hydrides* (Benjamin, New York 1965)

3.40 N. F. Mott, H. Jones: *The Theory of the Properties of Metals and Alloys* (Clarendon, Oxford 1936)

3.41 D. Zamir: Phys. Rev. **140**, A271 (1965)

3.42 K. F. Herzfeld, M. Goeppert-Mayer: Z. Phys. Chem. **26**, 203 (1934)

3.43 Y. Ebisuzaki, M. O'Keeffe: Prog. Solid State Chem. **4**, 187 (1967)

3.44 C. Kittel: *Introduction to Solid State Physics* 2nd ed. (Wiley, New York 1956)

3.45 J. M. Ziman: Adv. Phys. **13**, 89 (1964)

3.46 A. I. Gubanov, V. K. Nikulin: Sov. Phys. Solid State **7**, 2184 (1965)

3.47 J. J. Rush, R. C. Livingston, L. A. de Graaf, H. E. Flotow, J. M. Rowe: J. Chem. Phys. **59**, 6570 (1973)

3.48 A. Magerl, N. Stump, W. D. Teuchert, V. Wagner, G. Alefeld: J. Phys. C., Solid State Phys. **10**, 2783 (1977)

3.49 R. Yamada, N. Watanabe, K. Sato, H. Asano, M. Hirabayashi: J. Phys. Soc. Japan **41**, 85 (1976)

3.50 G. Verdan, R. Rubin, W. Kley: *Neutron Inelastic Scattering* Proc IAEA, Vienna (1968) p. 223

3.51 N. Stump, G. Alefeld, D. Tochetti: Solid State Commun. **19**, 805 (1976)

3.52 T. Springer: "Investigations of Vibrations in Metal Hydrides" in Topics Appl. Phys. Vol. 28 (Springer, Berlin, Heidelberg 1978) p. 75

3.53 N. A. Chernoplekov, M. G. Zemlyanov, V. A. Somenkov, A. A. Chertkov: Sov. Phys. Solid State **10**, 2783 (1977)

3.54 D. K. Ross, P. F. Martin, W. A. Oates, R. Khoda-Bakhsh: Z. Phys. Chem. **114**, 221 (1979)

3.55 R. Khoda-Bakhsh, D. K. Ross: J. Phys. F: Met Phys. **12**, 15 (1982)

3.56 J. J. Rush, N. F. Berk, A. Magerl, J. M. Rowe: Phys. Rev. B **37**, 7901 (1988)

3.57 J. J. Rush, H. E. Flotow: J. Chem. Phys. **48**, 3795 (1968)

3.58 R. Hempelmann, D. Richter, B. Stritzker: J. Phys. F: Met. Phys. **12**, 79 (1982)

3.59 W. Drexel, A. Murani, D. Tochetti, W. Kley, I. Sosnowska, D. K. Ross: J. Phys. Chem. Solids **37**, 1135 (1976)

3.60 D. G. Hunt, D. K. Ross: J. Less-Common Met. **49**, 169 (1976)

3.61 A. J. Maeland: J. Chem. Phys. **52**, 3952 (1970)

3.62 W. Wagener, P. Vorderwisch, S. Hautecler: Phys. Status Solidi B **98K**, 171 (1980)

3.63 P. Vorderwisch, S. Hautecler, H. Deckers: Phys. Status Solidi B **65**, 171 (1974)

3.64 J. J. Rush, H. E. Flotow, D. W. Connor, C. L. Thaper: J. Chem. Phys. **45**, 3817 (1966)

3.65 M. Saad: Atomkernenergie **17**, 281 (1971)

3.66 O. de Pous, H. M. Lutz: Second Int. Cong. Hydr. in Met., June 6–11, Paris 1977, P. 1E8

3.67 O. de Pous, H. M. Lutz: Proc. 2nd World Hydrogen Energy Conf., Zürich 1977 (Pergamon, Oxford 1978) p. 1597

3.68 H. Sugimoto, Y. Fukai: J. Phys. Soc. Japan **51**, 2554 (1982)

3.69 Y. Fukai: J. Less-Common Met. **101**, 1 (1984)

3.70 Y. Fukai: J. Less-Common Met. **172–174**, 8 (1991)

3.71 B. Baranowski, S. Majchrzak, T. Flanagan: J. Phys. F **1**, 288 (1971)

3.72 Y. Wong, F. B. Hill: AlChE **25**, 592 (1979)

3.73 T. Schober: In it Electronic Structure and Properties of Hydrogen in Metals, ed. by P. Jena, C. B. Satterthwaite (Plenum, New York 1983) p. 1

3.74 R. Lässer, K.-H. Klatt, P. Mecking, H. Wenzl: Ber. KFA Jülich **1800** (1982)

3.75 R. Lässer, T. Schober: J. Less-Common Met. **130**, 453 (1987)

3.76 I. R. Entin, V. A. Somenkov, S. Sh. Shilshtein: Sov. Phys. Solid State **16**, 1569 (1975)

3.77 R. Hempelmann, D. Richter, D. L. Price: Phys. Rev. Lett. **58**, 1016 (1987)

3.78 R. Hempelmann, D. Richter, D. L. Price: J. Less-Common Met. **130**, 203 (1987)

3.79 D. Klauder, V. Lottner, H. Scheuer: Solid State Comm. **32**, 617 (1979)

3.80 J. H. Wernick: In *Intermetallic Compounds*, ed. by H. Westbrook (Wiley, New York 1967) p. 197

3.81 W. B. Pearson: *Lattice Spacing and Structures of Metals and Alloys* (Pergamon, Oxford 1958)

3.82 L. Pauling: *The Nature of the Chemical Bond*, 3rd edn. (Cornell. Univ. Press, Ithaca 1960)

3.83 J. Tanaka, R. H. Wiswall, J. J. Reilly: Inorg. Chem. **17**, 498 (1978)

3.84 E. P. Magomedbekov, B. M. Andreev, A. V. Korolev: Zh. Fiz. Khim. **64**, 434 (1990) (in Russian)

3.85 B. Baranowski: "Metal-hydrogen Systems at High Hydrogen Pressures" in *Hydrogen in Metals*, ed. by G. Alefeld, J. Volkl, Topics Appl. Phys. Vol. 29 (Springer, Berlin, Heidelberg 1978) p. 157

3.86 J. F. Lynch, J. J. Reilly, J. Tanaka: "The Titanium–Molybdenum–Hydrogen System: Isotope Effects, Thermodynamics, and Phase Changes" in Trans. Met. Hydr., ed. by R. Bau (Am. Chem. Soc., Washington 1978) p. 342

3.87 M. Devillers, M. Sirch, R. D. Penzhorn: Z. Phys. Chem. NF **164**, 1355 (1989)

3.88 I. I. Wedernikova, E. P. Magomedbekov, B. M. Andreev, A. V. Krupentshenko, A. V. Korolev: Zh. Fiz. Khim. **65**, 1657 (1991) (in Russian)

3.89 B. Jungblut, G. Sicking: Z. Phys. Chem. NF **164**, 1177 (1989)

3.90 I. Jacob, A. Stern, A. Moran, D. Shaltiel, D. Davidov: J. Less-Common Met. **73**, 369 (1980)

3.91 F. T. Aldridge: J. Less-Common Met. **108**, 131 (1985)

3.92 G. D. Sandrock, J. J. Murray, M. L. Post, J. B. Taylor: Mat. Res. Bull. **17**, 887 (1982)

3.93 D. G. Westlake: J. Less-Common Met. **91**, 275 (1983)

3.94 R. Hempelmann, D. Richter, G. Eckold, J. J. Rush, J. M. Rowe, M. Montoya: J. Less-Common Met. **104**, 1 (1984)

3.95 D. G. Westlake: J. Mater. Sci. **19**, 316 (1984)

3.96 J. Shinar, D. Shaltiel, D. Davidov, A. Grayewski: J. Less-Common Met. **60**, 209 (1978)

3.97 R. L. Beck: "Investigation of Hydriding Characteristics of Intermetallic Compounds" (Lar-55-Denver Research Institute, 1961)

3.98 F. Pourarian, H. Fujii, W. E. Wallace, V. K. Sinha, H. K. Smith: J. Phys. Chem. **85**, 3105 (1981)

3.99 L. Pontonnier, S. Miraglia, D. Fruchart, J. L. Soubeyroux, A. Baudry, P. Boyer: J. Alloys Comp. **186**, 241 (1992)

3.100 J. H. N. van Vucht, F. A. Kuijpers, H. C. A. M. Bruning: Philips. Res. Rep. **25**, 133 (1970)
3.101 A. Drasner, Z. Blazina: J. Less-Common Met. **168**, 289 (1991)
3.102 F. Pourarian, W. E. Wallace: J. Solid State Chem. **55**, 181 (1984)
3.103 J. Bergsma, J. Goedkoop: Physica **26**, 744 (1960)
3.104 G. Boureau, O. Kleppa, P. Dantzer: J. Chem. Phys. **64**, 5247 (1976)
3.105 E. Glueckauf, G. P. Kitt: *Isotope Separation* Proc. Int. Symp. Amsterdam 1957 (North-Holland, Amsterdam 1958) p. 210
3.106 G. Sicking, P. Albers, E. Magomedbekov: J. Less-Common Met. **89**, 373 (1983)
3.107 P. Albers, Doctoral thesis, Münster, 1985
3.108 W. L. Whittemore, A. W. McReynolds: *Inelastic Scattering of Neutrons in Solids and Liquids* (IAEA, Vienna, 1961) p. 511
3.109 J. M. Rowe, J. J. Rush, H. G. M. Smith, H. E. Flotow: Phys. Rev. **B14**, 3630 (1976)
3.110 S. J. C. Irvine, D. K. Ross, I. R. Harris, J. D. Browne: J. Phys. F: Met. Phys. **14**, 2881 (1984)
3.111 E. Wicke, H. Brodowski: "Hydrogen in Palladium and Palladium Alloys", in *Hydrogen in Metals*, ed. by G. Alefeld, J. Völkl, Topics Appl. Phys. Vol. 29 (Springer, Berlin, Heidelberg 1978) p. 73
3.112 J. M. Nicol, J. J. Rush, R. D. Kelley: Phys. Rev. B **36**, 9315 (1987)
3.113 A. J. Renouprez, B. Fouillox, J. P. Candy: Surf. Sci. **83**, 285 (1979)
3.114 P. W. Albers, G. H. Sicking, D. K. Ross: J. Phys.: Condens. Matter **1**, 6025 (1989)
3.115 B. M. Andreev, A. N. Perevezentsev: Zh. Fiz. Khim. **55**, 2709 (1981) (English transl.: Russ. J. Phys. Chem. **55**, 1545 (1981))
3.116 V. Trentin, Ph. Brossard, D. Schweich: Chem. Eng. Sci. **48**, 873 (1993)

Chapter 4

4.1 A. S. Michaels: Ind. Eng. Chem. **44**, 1922 (1952)
4.2 A. M. Rozen: *Theory of Isotope Separation in Columns* (Atomizdat, Moskow 1960) p. 438 (in Russian)
4.3 A. M. Rozen (ed.): *Scaling in Chemical Technology* (Khimiya, Moscow 1980) p. 320
4.4 B. M. Andreev, A. S. Polevoi: Izv. Vyssh. Uchebn. Zaved. Khim. Khim. Tekhnol. **25**, 889 (1982) (in Russian)
4.5 B. M. Andreev, E. P. Magomedbekov, A. S. Polevoi: Mosk. Khim.-Tekhnol. im. D.I. Mendeleeva **156**, 24 (1989) (in Russian)
4.6 P. G. Romankov, N. B. Rashkowskaya, W. F. Frolov: *Mass-Transfer Processes in the Chemical Industry* (Khimiya, Leningrad 1975) p. 192 (in Russian)
4.7 M. E. Aerov, O. M. Todes, D. A. Narinskii: *Apparatuses with Immobile Granular Layer* (Khimiya, Leningrad 1979) p. 176 (in Russian)
4.8 P. G. Romankov, W. N. Lepilin: *Continuous Adsorption of Gases and Vapours* (Khimiya, Leningrad 1968) p. 228 (in Russian)
4.9 M. S. Safonov, W. K. Shirjaev, W. I. Gorshkov: Zh. Fiz. Khim. **44**, 975 (1970) (in Russian)
4.10 W. I. Gorshkov, M. S. Safonov, N. M. Woskresenskii: *Ion Exchange in Counter-Current Columns* (Nauka, Moscow 1981) p. 224
4.11 W. K. Shirjaev, M. S. Safonov: Theor. Osnovi Khim. Tekhnol. **3**, 922 (1969) (in Russian)
4.12 K. Soga, H. Imamura, S. Ikeda: Nippon Kagaku Kaishi **9**, 1304 (1977)
4.13 B. M. Andreev, A. N. Perevezentsev, V. V. Shitikov: Zh. Fiz. Khim. **55**, 1993 (1981)
4.14 G. Sicking, P. Albers, E. Magomedbekov: J. Less-Common Met. **89**, 373 (1983)

4.15 K. I. Blank, E. P. Magomedbekov, A. V. Krupentshenko: Tr. Mosk. Khim.-Teknol. Inst. im. D.I. Mendeleeva **147**, 59 (1987) (in Russian)

4.16 L. Schlapbach: J. Phys. F. **10**, 2477 (1980)

4.17 H. C. Siegman, L. Schlapbach, C. R. Brundle: Phys. Rev. Lett. **40**, 785 (1979)

4.18 L. Schlapbach, A. Seiler, F. Stucki, H. C. Siegman: J. Less-Common Met. **73**, 145 (1980)

4.19 S. H. Overbury, P. A. Bertrand, G. A. Somorjai: Chem. Rev. **75**, 547 (1975)

4.20 B. M. Andreev, E. P. Magomedbekov, Yu. S. Pak, A. A. Firer: Tr. Mosk. Khim.-Tekhnol. Inst. im. D.I. Mendeleeva **147**, 59 (1987) (in Russian)

4.21 E. P. Magomedbekov, B. M. Andreev, A. W. Korolev: Zh. Fiz. Khim. **64**, 434 (1990) (in Russian)

4.22 I. I. Wedernikova, E. P. Magomedbekov, B. M. Andreev, A. V. Krupentshenko, A. V. Korolev: Russ. Zh. Fiz. Khim. **65**, 1657 (1991) (in Russian)

4.23 B. Jungblut, G. Sicking: J. Phys. Chem. NF **164**, 1177 (1989)

4.24 B. M. Andreev, G. K. Boreskov, C. Chang-tsun, V. M. Tsionski: Kinet. Katal. **7**, 470 (1966) (English transl.: Kinet. Catal. **7**, 416 (1966))

4.25 B. M. Andreev, A. N. Perevezentsev, V. I. Yasenkov: Zh. Fiz. Khim. **55**, 423 (1981) (English transl.: Russ. J. Phys. Chem. **55**, 232 (1981))

4.26 D. H. W. Carstens: J. Phys. Chem. NF **164**, 1185 (1989)

4.27 D. H. W. Carstens, P. D. Encinias: J. Less-Common Met. **172–174**, 1331 (1991)

4.28 G. W. Foltz, C. F. Melius: J. Catal. **108**, 409 (1987)

4.29 B. M. Andreev, A. S. Polevoi, O. V. Petrenik: At. Energy **40**, 431 (1976) (English transl.: Sov. At. Energy **40**, 516 (1976))

4.30 J. J. Scholten, J. A. Konvalinka: J. Catal. **5**, 1 (1966)

4.31 E. Wicke, H. Brodowsky: In Topics Appl. Phys. Vol. 29 (Springer, Berlin, Heidelberg 1978) p. 73

4.32 K. I. Blank, E. P. Magomedbekov, A. V. Krupentshenko: Mosk. Khim.-Tekhnol. Inst. im. D.I. Mendeleeva **130**, 70 (1984) (in Russian)

4.33 R. Hempelmann: J. Less-Common Met. **101**, 69 (1984)

4.34 L. Belkbir, N. Gerard, A. Percheron-Guegan, J. C. Achard: Int. J. Hydrogen Energy **4**, 541 (1979)

4.35 H. Uchida, H. Uchida: J. Less-Common Met. **89**, 495 (1983)

4.36 T. Gamo, Y. Moriwaki, N. Yanagihara, T. Iwaki: J. Less-Common Met. **89**, 495 (1983)

4.37 D. Richter, R. Hempelmann, L. A. Vinhas: J. Less-Common Met. **88**, 353 (1982)

4.38 H. Buhl, S. Will: "Sorbent for Hydrogen Storage" Patent 4110425 USA, Int. Cl.³. CO1B6/24 (1978)

4.39 H. Buhl, S. Will: "Sorbent for Hydrogen Storage" Patent 2550584 BRD, Int. Cl.³. CO1B6/24 (1977)

4.40 S. S. Batsanov, L. I. Kapaneva, E. V. Lasarev: Zh. Neorg. Chem. **28**, 1063 (1983) (in Russian)

4.41 E. Henley, E. Johnson: *Radiation Chemistry* (Russian transl.: Atomisdat, Moscow 1974); E. Henley, E. Johnson: *The Chemistry and Physics of High Energy Reactions* (Oxford University Press, Oxford 1969) p. 415

4.42 P. S. Rudman, G. D. Sandrock, P. D. Goodell: J. Less-Common Met. **89**, 437 (1983)

4.43 I. A. Adrova, N. I. Bessonov: *Polyimides – A Class of Heat-Resistant Polymers* (Nauka, Leningrad 1968) p. 210 (in Russian)

4.44 B. M. Andreev, E. P. Magomedbekov, Yu. S. Pak, G. N. Shwedova: Vyssh. Uchebn. Zaved. Khim. Khim. Tekhnol. **29**, 54 (1986) (in Russian)

4.45 J. J. Scheridan, F. G. Eisenberg, E. J. Greskovich, G. D. Sandrock, E. L. Huston: J. Less-Common Met. **89**, 447 (1983)

4.46 M. Ron, D. Gruen, M. Mendelsohn, I. Sheft: J. Less-Common Met. **74**, 445 (1980)

4.47 H. Ishikawa, K. Oguro, A. Kato, H. Suzuki, E. Ishii: J. Less-Common Met. **107**, 105 (1985)

4.48 E. Tuscher, P. Weinzierl, O. Eder: Int. J. Hydr. Energy **8**, 199 (1983)

4.49 M. Ron: "Method for Preparing Improved Porous Metal–Hydride Compacts and Apparatus Therefor" Patent 2126206. UK. Int. Cl^3. C01B6/24 (1983)

4.50 Y. Suzuki, Z. Ogama: "Hydrogen Occlusion Material" Patent 59-73401. Japan. Int. Cl^3. C01B6/24 (1984)

4.51 G. D. Sandrock, P. D. Goodell: J. Less-Common Met. **73**, 61 (1980)

4.52 K. Suzuki: J. Less-Common Met. **89**, 183 (1983)

4.53 B. M. Andreev, A. N. Perevezentsev, Y. N. Pisarev, S. M. Ivanov: Izv. Akad. Nauk SSSR, Neorg. Materialy **23**, 233 (1987) (English transl.: Inorg. Materials **23**, 204 (1987))

4.54 B. M. Andreev, E. P. Magomedbekov, G. N. Shwedova, I. N. Levin: Zh. Fiz. Khim **61**, 1827 (1987) (in Russian)

4.55 Y. S. Lezin, M. M. Dubinin: Dokl. Akad. Nauk. SSSR **171**, 382 (1966) (in Russian)

Chapter 5

5.1   E. G. Glueckauf, G. P. Kitt: Angew. Chem. **69**, 567 (1957)

5.2   E. Glueckauf, G. P. Kitt: *Isotope Separation*, Proc. Int. Symp. Amsterdam, 1957 (North-Holland, Amsterdam 1958) p. 20

5.3   J. E. How: Science **161**, 464 (1968)

5.4   S. Tistchenko, G. Dirian: Bull. Soc. Chim. France **1**, 16 (1970)

5.5   B. M. Andreev, A. N. Perevezentsev, V. I. Yasenkov: Zh. Fiz. Khim. **55**, 423 (1981) (English transl.: Russ. J. Phys. Chem. **55**, 232 (1981))

5.6   B. M. Andreev, A. S. Polevoi, A. N. Perevezentsev: Radiokhimia **28**, 489 (1986) (in Russian)

5.7   F. Botter, D. Leger, R. Darras: Bull. Inform. Sci. Technol. CEA **183**, 25 (1973)

5.8   F. Botter, J. Gowman, J. L. Hemmerich, B. Hircq, R. Lässer, D. Leger, S. Tistchenko, M. Tschudin: Fus. Technol. **14**, 562 (1988)

5.9   B. Hircq: Fus. Technol. **14**, 424 (1988)

5.10 M. C. Embury, R. E. Ellefson, H. B. Melke, W. M. Rutherford: Fus. Technol. **21**, 960 (1992)

5.11 K. Weaver, C. E. Hamrin, Jr.: Chem. Eng. Sci. **29**, 1873 (1974)

5.12 L. H. Shendalman, J. E. Mitchell: Chem. Engng. Sci. **27**, 1449 (1972)

5.13 Y. W. Wong, F. B. Hill, Y. N. I. Chan: Sep. Sci. Technol. **15**, 423 (1980)

5.14 M. S. Ortman, L. K. Heung, A. Nobile, R. L. Rabun III: J. Vac. Sci. Technol. **A8**, 2881 (1990)

5.15 A. S. Horen, M. W. Lee: Fus. Technol. **21**, 282 (1992)

5.16 D. Basmadjian: Can. J. Chem. Engng. 269 (Dec. 1963)

5.17 A. Clayer, L. Agneray, G. Vandenbusche, P. Petel: Z. Anal. Chem. **236**, 240 (1968)

5.18 B. M. Andreev, A. S. Polevoi: Dokl. Akad. Nauk Grus. SSR **7**, 181 (1981) (in Russian)

5.19 B. M. Andreev, Ya. D. Zelvenskii, S. G. Katal'nikov: *Heavy Isotopes of Hydrogen in Nuclear Technology* (Energoatomisdat, Moskow 1987) p. 456 (in Russian)

5.20 B. M. Andreev, G. K. Boreskov: Zh. Fiz. Khim. **38**, 115 (1964) (in Russian)

5.21 B. M. Andreev, A. S. Polevoi: Zh. Fiz. Khim. **56**, 349 (1982) (in Russian)

5.22 B. M. Andreev, A. N. Perevezentsev, I. L. Selivanenko: *Isotope Separation and Chemical Exchange Uranium Enrichment*, Proc. Int. Symp., Tokyo, October 29–November 1, 1990, eds. Y. Fujii, T. Ishida, K. Takeuchi, Bull. Research Lab. Nucl. Reactors **1**, 1 (1992)

5.23 A. N. Perevezentsev, B. M. Andreev, I. L. Selivanenko, I. A. Yarcho: Fus. Engng. Design **18**, 39 (1991)

5.24 K. Bier: Chem.-Ing.-Techn. **28**, 625 (1956)

5.25 A. M. Rosen: *Theory of Isotope Separation in Columns* (Atomisdat, Moscow 1960) (in Russian)

5.26 B. M. Andreev, G. K. Boreskov, S. G. Katal'nikov: Khim. Prom. **6**, 19 (1961) (in Russian)

5.27 B. M. Andreev, A. S. Polevoi: Izv. Vyssh. Uchebn. Zaved. Khim. Khim. Tekhnol. **25**, 889 (1982) (in Russian)

5.28 E. Ewans: *Tritium and Its Compounds* (Russian transl.: Atomisdat, Moscow 1970)

5.29 Ph. Pautron, J. P. Arnauld: Trans. Am. Nucl. Soc. **20**, 202 (1975)

5.30 B. Misra, V. A. Maroni: Nucl. Technol. **35**, 40 (1977)

5.31 A. Busigin, S. K. Sood, O. K. Kveton, R. H. Sherman, J. L. Anderson, L. J. Wittenberg: *ITER Isotope Separation System Conceptual Design Description*, Revision 1.0, September 20, 1990

5.32 R. D. Penzhorn, J. Anderson, R. Haange, B. Hircq, A. Meikle, Y. Naruse: Fus. Engng. Design **16**, 141 (1991)

5.33 B. M. Andreev, A. V. Kruglov, A. N. Perevezentsev, M. B. Rozenkevich, Z. V. Ershova, V. L. Zverev, A. V. Kapyshev: Tr. Mosk. Khim.-Tekhnol. Inst. im. D.I. Mendeleeva **156**, 49 (1989)

5.34 B. M. Andreev, M. V. Karpov, A. N. Perevezentsev, M. B. Rozenkevich, Yu. A. Sakhoravskii: Vodorodnaya Energetika i Tekhnolog. **1**, 57 (1992) (in Russian)

5.35 B. M. Andreev, Z. V. Ershova, V. I. Zverev, A. V. Kapyshev, A. N. Perevezentsev, M. B. Rozenkevich: Voprosy Atomnoy Nauki i Tekhniky. Termojadernly Sintez **2**, 55 (1990) (in Russian)

List of Symbols

Me	hydride-forming metal
$\Delta \bar{G}_H$	change of partial molar isobaric-isothermal potential (change in GIBBS free energy) on hydrogen dissolving in metals
$\Delta \bar{H}_H$	change of partial molar enthalpy
$\Delta \bar{S}_H$	change of partial molar entropy
S^{ci}	ideal configurational entropy
n	ratio of hydrogen and metal atoms (n_α at the boundary of α and $\alpha + \beta$; n_β at the boundary of $\alpha + \beta$ and β; and n_s the greatest possible value of n)
K_S	Sieverts constant
α	symbol for solid hydrogen solution in metals; separation factor
β	symbol for hydride phase
$\Delta H_{\alpha - \beta}$	change of enthalpy in the act of hydride formation
$\Delta S_{\alpha - \beta}$	change of entropy in the act of hydride formation
H_α, H_β	partial molar enthalpies of hydrides in α and β phases
S_α, S_β	partial molar entropies of hydrides in α and β phases
θ_H	the Einstein characteristic temperature
ω_H	local mode frequency of hydrogen atoms in the cyrstal lattice of metals or IMC
$\hbar$	the Planck constant, $\hbar = h/2\pi$
k	the Boltzmann constant
ε	energy of local modes of hydrogen in the crystal lattice of metals and IMC; enrichment factor in Chap. 4
μ_{H_2}	chemical potential of gaseous hydrogen
μ_H	chemical potential of hydrogen in the crystal lattice of a metal
$\bar{V}_H$	volume occupied by 1 g atom of hydrogen in the crystal lattice
u	reduced temperature, $u = \hbar\omega/kT$

N_g, N_s	the number of hydrogen moles in gas and solid phases, respectively
F	exchange degree
r	observed rate constant
τ	time
$\tau_{0.5}$	half-exchange time
S_0	fraction of free section of column
S	contact surface area of phases; column section
S_{sp}	specific surface of Me or IMC
x_∞, y_∞	equilibrium concentration of heavy isotope in hydride and gas phases, respectively
g	mass of solid phase of Me or IMC
K_{0g}	mass-transfer coefficient for gas phases
β_g, β_s	mass-exchange coefficients in gas and solid phases
D_{eff}	effective coefficient of axial diffusion
D	coefficient of molecular diffusion of hydrogen
w	linear gas velocity
w_T	velocity of temperature-zone movement
m	slope of the equilibrium line ($m = dx/dy$)
h_{0g}, h_{0s}	height of transfer unit (HTU) for gas and solid phases, respectively
h_g, h_s, h_{ad}, h_p	components of HTU caused by mass-exchange in gas and solid phase by axial diffusion in the gas, and diffusion in pores
G_{sp}, L_{sp}	specific molar flows (molar flow densities) of hydrogen in gas and hydride phase, G_{sp} is also called load (of the column)
λ	flow ratio ($\lambda = G_{sp}/L_{sp}$)
S_{fr}	free section of the column
ϱ	hydrogen density
a	contact surface of phases per unit of column volume
M	molecular mass of hydrogen
Nu_g	Nusselt number
Re_g	Reynolds number
Pr_g	Prandtl number
$m_{A,B}$	mass of light or heavy hydrogen isotope
$\overline{H}_i^{Me}$	enthalpy of metal matrix in α or β phase
$P_{H_2}, P_{D_2}, P_{T_2}$	equilibrium pressures of protium, deuterium, and tritium, respectively

A	light isotope
B	heavy isotope; transition metal
$P_{H_2}^A$	equilibrium pressures determined by sorption isotherm
$P_{H_2}^D$	equilibrium pressures determined by desorption isotherm
x, y	atomic fraction of heavy hydrogen isotope (deuterium or tritium) in hydride and gas phases, respectively
P_1, P_2	equilibrium pressure of pure components over hydride phase
α_{A-B}^0	separation factor at equal ratio of isotopes A and B in the gas phase
α_{AB}, α_{BA}	separation factors in the range of low content of light and heavy isotopes, respectively
$\bar{\alpha}$	effective separation factor
ϕ	fraction of hydrogen atoms retained in the α-phase of the two-phase region with respect to the overall amount of hydrogen atoms in the range of $\alpha-\beta$ transition
K_{A-B}	equilibrium constant of the reaction of gaseous hydrogen isotope exchange with hydride phases of Me or IMC
K_{AB}	equilibrium constant of the homomolecular isotope exchange (HMIE) reaction of hydrogen $A_2 + B_2 \leftrightarrow 2AB$
Z_{A_2}, Z_{B_2}	partition functions of hydrogen molecules containing light and heavy isotope, respectively
$Z_{A(Me)} Z_{B(Me)}$	partition functions of hydride phases of metal or IMC including light and heavy hydrogen isotopes, respectively
σ	symmetry number of molecules
m_A, m_B	masses of hydrogen atoms A and B, respectively
K	degree of separation
N	number of theoretical plates of separation
N_{0g}, N_{0s}	number of transfer units for gas and solid phases, respectively
R	rate of the isotope exchange reaction
d_e	equivalent size of channels in a granular bed
μ	gas viscosity
a_{gr}	geometrical surface area of grains of Me or IMC per unit volume of the column
d_{gr}	diameter of spherical grains
D_e	effective coefficient of hydrogen diffusion in grains of Me or IMC
E_H	hydrogen content per unit of sorbent volume
h_e	height equivalent of a theoretical plate of separation (HETP)

ξ	form coefficient
Z	height of the sorbent bed in a column
H	height of the separation part of a column
ϕ	relative withdrawal
E, E_{m}	extraction degree
P	plant productivity
B, P	flows of withdrawal and waste

List of Materials

(Excluding those given in Tables and Figures)

Subject Index

Springer Tracts in Modern Physics

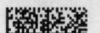